The Whale Research Assistant Program

The Azores Archipelago Version

Student Study Book and Manual

This manual has been published by
Conservation Diver Ltd

© 2008 – 2023 Conservation Diver, all rights reserved

Written by George Bevan & Joana Vaz Pereira

With contributions by Elouise Haskin, Chad Scott & Ruth Pontin
Illustrations by Pau Urgell Plaza

This book may not be reproduced, copied, or replicated without explicit written permission from the creator or Conservation Diver Ltd. Pt.

This manual only provides a background and a broad outline for conducting the Whale Research Assistant Program; this does not serve as an alternative to learning the course with an instructor. Anybody wishing to conduct the Whale Research Assistant Program should do so only under the close supervision of a trained professional to prevent harm to the environment or person. The writer of this guide and Conservation Divers Ltd. Prt. are not responsible for any injuries or problems sustained while participating in the program. SCUBA diving and other water activities can be dangerous and are done at your own risk.

ISBN: 978-1-7326925-4-1

Contents

Whale Research Assistant Program — 4
Chapter 1: Cetacean Ecology — 9
What are Cetaceans — 10
Evolutionary History of Cetaceans — 11
Mysticetes — 13
Odontocetes — 14
Cetacean Occurrence in the Azores — 15
The Value of Cetaceans — 19
Communication — 22
Diving Physiology — 25
Feeding — 26
Identifying Cetaceans — 29
Cetacean Behavior — 33
Guidelines to Approaching Cetaceans — 38
The Role of the Lookout — 40
Observation Slates — 40
Identification Slates — 41
Chapter 2: Baleen Cetaceans — 44
Blue whale — 45
Fin whale — 49
Sei whale — 53
Humpback whale — 56
Minke whale — 61
Bryde's whale — 64
Chapter 3: Research Focus Cetacean — 68
Sperm Whale — 69
Chapter 4: Toothed Cetaceans — 78
Orca — 79
False Orca — 82
Short-finned Pilot Whale — 84
Bottlenose Dolphin — 86
Risso's Dolphin — 88
Common Dolphin — 89
Striped Dolphin — 91
Atlantic Spotted Dolphin — 92
Chapter 5: Research & Conservation — 95
Cetacean Research — 96
Cetacean Conservation — 103
Citations — 106
Whale Hotspots of the World — 107

Acknowledgements

This manual would not have been possible without the tireless efforts of the team at Dive Azores who formed the baseline study on which this course was written. Their work and experience over the last decade has provided Conservation Diver with the key sources of information needed to make citizen science more widely available to the whale watching community.

The creation of this course would also not have been possible without the wider support network that is at the heart of Conservation Diver.

The Whale Research Assistant Program

This course is designed to give both whale enthusiasts and conservationists the platform to learn about whale ecology, execute photo-ID research, collate and manage data, understand other research methods and develop the skills to work in a whale watching and research station. By increasing awareness and involvement, we can decrease our impact on the oceans and whale populations to help provide solutions to protect and restore populations in the Azores Archipelago region and around the world. The most successful projects to conserve the world's oceans have stemmed from passionate individuals and action orientated groups and communities- not governments or policy makers. We hope to continue in this tradition and do what we can to protect the environment which we are part of.

After completing this course, students will know how to perform the research methods of the Whale Research Assistant program and how to continue to help with data collection both within and outside the region. From the program they will gain knowledge about whale ecology, the difference between species with regards to anatomy, physiology and social structures; how to identify different species and individuals, as well as the current threats, conservation and research efforts taking place to protect the remaining populations. Students will gain experience collating data within a busy research station and develop the required skills both on land and sea to complete such activities. Students who complete this training can then volunteer to assist with data collection and management at a research station or apply the skills they have learnt to different areas in the world through other citizen science programs.

Prerequisites

This program is a non-diving course. Hence students are not required to hold an advanced diving qualification in order to take part. (Please note that this is a prerequisite for many other Conservation Diver courses). We do however require that participants adhere to the following prerequisites for beginning this recognition program:

- Be at least 16 years of age before program commencement
- Review all safety procedures and understand the particular risks of the areas you will be working in
- Sign any locally required health and liability wavers

Its is also **highly recommended** that students obtain a high quality DSLR or mirrorless camera with a powerful zoom lens (minimum 300 mm) in order to fully contribute to the photo identification elements of the course. All instructors are in possession of such photography equipment which will be used for classroom evaluation and data input.

Students will be given a full packing list of items to bring with them on the program.

Course Outline

The following course outline is based on a 14+ day program, which is believed to be the minimum amount of time to properly teach and give the experience necessary for certification. It is, however, highly recommended that students opt to stay on for further weeks in order to refine their skills and to ensure that they meet all the requirements and standards for certification. The exact daily schedule will be adapted to fit in with weather conditions and other operational constraints at the research station.

Lectures

The program contains 6 lectures that will be delivered in whole, or in parts, depending on group progression and operational constraints. They consist of the following:

- Orientation
- Cetacean Ecology and Life History
- Baleen Cetaceans
- Research Focus Cetacean
- Toothed Cetaceans
- Research & Conservation

100% attendance is required as part of the requirements of this program to gain certification.

e-Learning

Students are required to complete the corresponding e-Learning, which will be evaluated with your instructor to ensure sufficient understanding.

Examination

In order to gain certification in the course, students must complete a multiple-choice exam. The pass rate is 80%.

Whale Monitoring Excursions

Each week, students will partake in a minimum of 4 whale monitoring excursions at sea. Often this number is higher depending on other work and weather conditions. Excursions can be anywhere between 2 and 6 hours, and so students must be prepared to be at sea for large parts of the day.

Whale History and Local Industry

During the program students will be taken to the local whaling museum to learn the history of the whaling industry in the Azores. The remnants of the whale industry will also be seen as students visit the land lookouts, or 'Vigias', who now work with the whale watching centers. They will also learn about the craft of scrimshaw and visit one of the largest collections in the world.

Standards and Requirements

In order to complete the program and achieve certification under Conservation Diver, students must meet all of the standards and requirements listed below:

	Requirements	Standard
Whale Research Assistant	<ul><li>Attend all 6 Lectures</li><li>Attend a minimum of 4 boat excursions per week</li><li>Complete all e-Learning</li><li>Complete Multiple Choice Exam</li><li>Pass Group Presentation task</li><li>Manage data and Input into the Database</li><li>Demonstrate proper techniques to gather data and take usable photographs</li><li>Demonstrate a good level of boat management skills and competency</li><li>Maintain a professional standard as ambassadors of the program</li></ul>	<ul><li>Learn about cetacean biology</li><li>Learn the differences between the cetaceans found in the region of study, including but not limited to: anatomy, feeding, society and culture and reproduction</li><li>Learn how to take a cetacean survey in the field and record data</li><li>Learn the methods to gather data, the camera skills required and the process of Photo-ID</li><li>Learn the current threats facing cetaceans and the conservation efforts being made to protect them.</li><li>Learn how to participate in the ongoing protection of cetaceans</li><li>Learn the scientific cetaceans research taking place in the around the globe</li><li>Learn about other cetacean hotspots around the globe and the other species found there</li><li>Understand the role of a look-out</li><li>Be able to identify the different whales and dolphins found in the region of study but recognizing features such as their blow, surfacing/diving, dorsal fin, sizes, and other specific characteristics</li><li>Be able to collate data, manage and catalog the research database</li><li>Understand how to analyse the data and conduct scientific research of the outputs</li><li>Learn basic boat skills and assist with boat duties including boat check and post trip clean down</li></ul>

The Conservation Diver Instructor withholds the right to deny certification if any of the above has not been met satisfactorily.

Certification Process

Once the instructor has determined that the standards outlines above have been met, and the requirements have been fulfilled, then the student will be signed off and certified. The certification is then processed at Conservation Diver, who will distribute the recognition card and supporting materials.

Please ask your instructor or check our website for information about our other courses and opportunities within Conservation Diver.
Should you have any more questions or queries about the certification process please reference the FAQs page on our website or email us:

 www.ConservationDiver.com

 info@conservationdiver.com

Cetaceans in the Azores Archipelago Region

The Azores are an autonomous region of Portugal in the middle of the Atlantic Ocean and are one of the world's best destinations for encountering whales and dolphins in the wild. Attracted by the deep waters that occur close to shore and an abundance of food, more than 25 species of whales and dolphins are reported to the Azorean waters: about 30% of all the known cetacean species.

Considered a sperm whale "hotspot", it is also on the route of some migrating species, such as the blue, fin and sei whales. Few people know the Azores as a blue whale destination, but in fact it is one of the best places in the North-East Atlantic to get up close with the biggest animals on Earth.

The Whale Research Assistant program operates on Faial Island, in conjunction with Dive Azores: a research station that combines a touristic whale watching operation with research- involving guests and volunteers in a philosophy of sustainable, research-led, eco-tourism.

The research station is run by a team of marine biologists who have been conducting whale photo-ID studies in the Azores for over 10 years.

The focus of the study is mainly on the individual identification of blue *(Balaenoptera musculus)* and sperm *(Physeter macrocephalus)* whales. It consists of two research periods: the spring months- mainly for blue whale research (fin, sei, humpback and sperm whales as well) and the summer, where efforts are focused on the individual recognition of sperm whales.

Long-term photo-ID studies can provide important information on whale movements, life history and population size, and it is one of the most valuable whale research techniques. The Whale Research Assistant Program gives you the chance to experience field research and to meet whales and dolphins in their natural realm, whilst learning about cetacean biology, conservation, species identification, photo-ID and behavioral data collection techniques.

The Azores has managed its growth in tourism well, helped by its remote location, seasonality and limited infrastructure. The whale observation operators must adhere to strict regulations and codes of conduct. However, there is still room for further revisions, and an adaptive management plan of regulations is able to grow alongside scientific research and observations.

Animals of the Azores

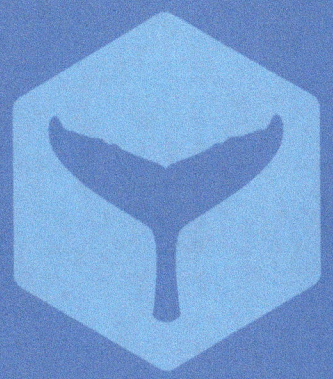

Chapter 1
Cetacean Ecology

"Ecology is the study of organisms and their interactions with each other and their environment."

In order to monitor the whales, you must first have an understanding about how they function and how they live.

Chapter 1
Whale Ecology

What are Cetaceans?

Cetaceans are aquatic mammals constituting the infraorder Cetacea, which consists of all whales, dolphins and porpoises. They are the biggest animals that have ever lived and are found across all oceans. There are 90 recognized species, all of which are marine, except 5 extant species of river dolphin. The number of species has the tendency to change, or varies across different literature, since the science is not static. New species are being discovered, as well as existing species or subspecies being split or revised in taxonomic terms.

Cetaceans range in size from Maui's dolphin, the smallest at 1 m (3 ft 3 in) in length and 50 kg (110 lbs) to the largest- the Blue whale- reaching lengths of 30 m (98 ft) weighing in at 190 tonnes (209 tons). Cetaceans share the common attributes of having no hindlimbs and flattened forelimbs that form flippers, boneless tail flukes, a nasal opening on the top of the head for ease of breathing, a layer of blubber to maintain body heat and streamlined bodies to reduce drag.

Modern Cetacea are divided into two parvorders:

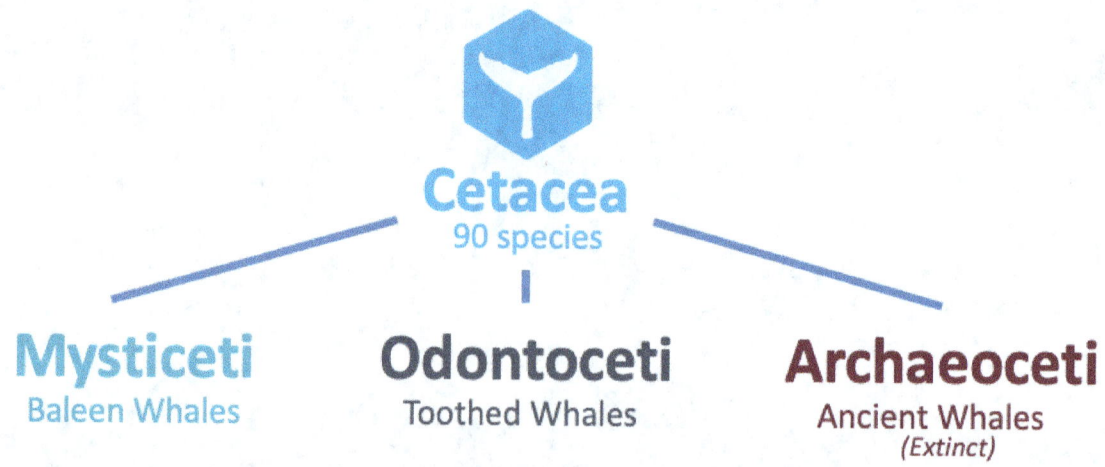

Mysticeti are the baleen whales and Odontoceti are the toothed whales, dolphins and porpoises. Archaeoceti are the ancient ancestors of modern whales living around 55 – 45 million years ago, becoming extinct in the early Miocene.

They diverged from the even-toed ungulate (hooved) ancestors of cetaceans known as Pakicetus, evolving in the shallow waters that separated India from Asia. They were semi-aquatic, however some species adapted to be fully oceanic. Around 34 million years ago Mysticeti split from Odontoceti, developing baleen for filter feeding food. Today the closest living relative of cetaceans are hippopotamuses.

Pakicetus (early Eocene 49–48 million years ago) about 1.75 m (6 ft) in length with functioning forelimbs and hindlimbs for walking. Foraged in streams and had no tail fluke.

Evolutionary History of Cetaceans

Cetaceans evolved rapidly from land to sea over a period of 14 million years, in a response to environmental changes. Pakicetus, which existed in the early Eocene, 48 million years ago, was a semi-aquatic animal found foraging in and around streams; eating plants at the water's edge, which gave it the opportunity to hide from predators in the shallows. It was about 1.75 m (6 ft) in length with functioning forelimbs and hindlimbs for walking, but it had no tail fluke.

Over the next 1 million years, its descendants spent more and more time in and around the shallows, which saw the emergence of the Ambulocetus, an alligator like animal with a larger powerful tail and shorter legs. It was larger, at 4.15 m (14 ft) in length, and fed in the water.

The long-snouted and otter-like Remingtonocetids, such as the Kutchicetus, appeared next, living throughout the near-shore environments- from the marsh lands to shallow waters. At about the same time appeared an even more aquatically adapted group known as the Protocetids. Protocetids,

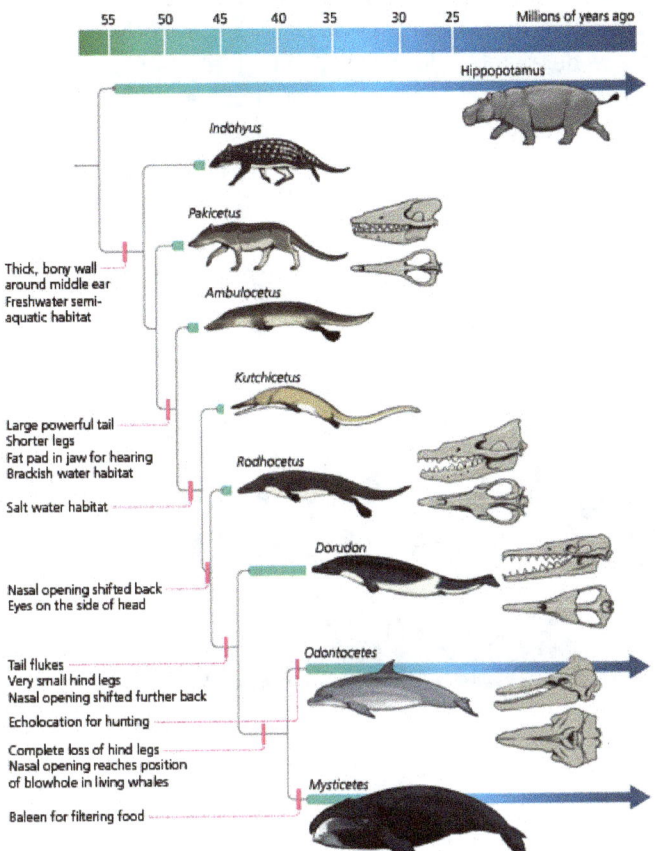

Whale phylogeny from The Tangled Bank, Carl Zimmer

like the Rodhocetus, were nearly entirely aquatic.

After spending a greater time in the water, the body evolved to be more streamlined and adapted to swimming, as well as blubber replacing fur coats for greater warmth and streamlining. Here, 46 million years ago, the Maiacetus emerged: a sealion-like animal, only spending a small amount of time on land, which meant its forelimbs functioned for walking and swimming, with larger hindlimbs.

Gradually the tail became bigger and stronger for swimming, which elongated the trunk of the animal and so 9 million years later, the Dorudon came into existence, evolved to be fully aquatic. At 4.5 m (15 ft), the Dorudon's forelimbs became flippers; hindlimbs were considerably regressed and the tail fluke evolved. The nasal openings shifted further back on the skull, reaching the same position we find the blowhole of living whales, so that they could breathe easily without the need to tilt their heads while swimming. Their eyes moved to the side of the head and eventually Dorudons completely lost their hindlimbs.

Then, around 34 million years ago, the Llanocetus evolved, whose teeth were reduced and showed the first evidence of baleen for filter feeding. This caused the splitting into the parvorders: Mysticeti and Odontoceti, which consist of all the modern-day whales, dolphins and porpoises- split across 14 families (although the Yangtze river dolphin is now believed to be extinct).

Many species then evolved during the progression towards the modern-day whale taxonomy. Late in the Oligocene, c.25 million years ago, evolved the Squalodon. These cetaceans had large and slightly convex depressions for air sacs on their elongated rostrum, the beaklike projection on their heads. This is associated with the presence of a melon, presumed to be the first appearance of echolocation.

Squalodon

Janjucetus

Around the same time, basal baleen whales evolved- specifically the Janjucetus- who were macrophagous feeders without baleen; possessing sharp teeth for gripping and tearing prey. It is believed therefore that they had similar hunting behaviors to the extant leopard seal (*Hydrurga leptonyx*) and shearing dentition similar to many archeocyte cetaceans. These basal Mysticetes did not use echolocation or filter feeding, hence we know that the evolution of bulk filter feeding for Mysticetes was a gradual process and that some archaic Mysticetes were quite unlike their modern relatives.

Around 9.8 to 8.9 million years ago, during the Miocene, evolved the raptorial sperm whale: the ancestor of modern sperm whales. This family of whales had very large teeth in both jaws, that were deeply rooted and interlocking, helping to keep hold of struggling prey. They hunted large marine mammals including other whales, similarly to the modern-day orca, and probably competed for food with the extinct giant shark, Megalodon. They were large whales, the largest species being the *Livytan melvillei* (named after the author of Moby Dick, Herman Melville) measuring 13.5–17.5 m (44–57 ft) in length. This species was also distinguished from other raptorial sperm whales by the basin on its skull, spanning the entire length of the rostrum, which contained the spermaceti organ: the origins of echolocation.

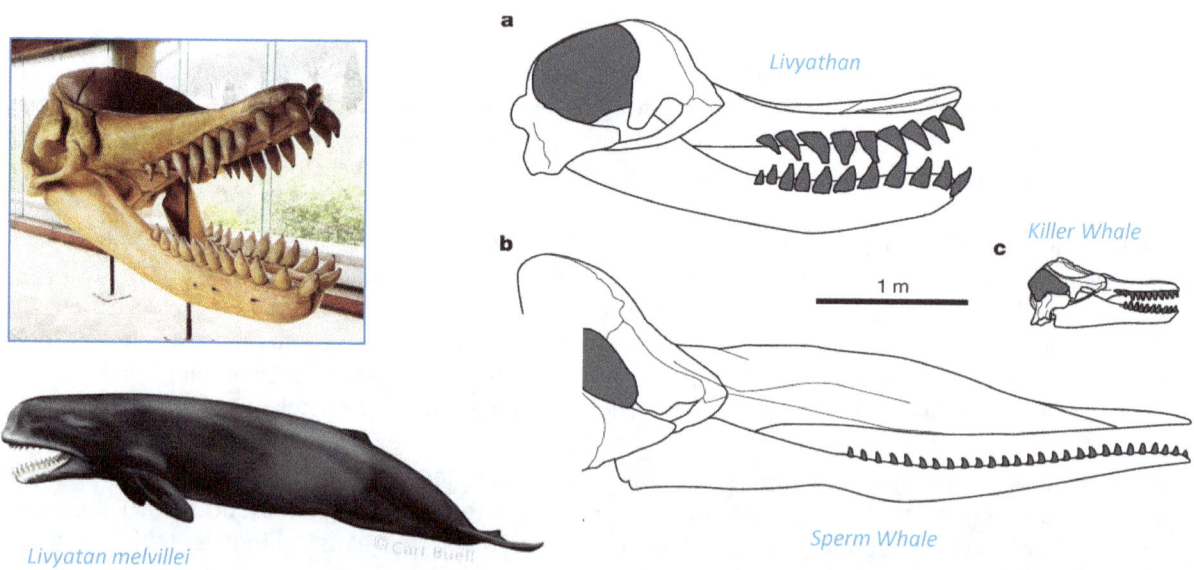

Livyatan melvillei

Livyathan

Killer Whale

Sperm Whale

The early representatives of modern whales appeared over the next million years, ultimately giving rise to forms as diverse as the pink Amazon river dolphin and the gigantic blue whale.

The Whale Research Assistant Program, 1st Edition

Mysticetes

Mysticetes are commonly known as baleen whales. Instead of teeth, they have tightly packed, comb-like, keratinous plates hanging down from the upper jaw. These baleen plates act as giant sieves, allowing for obligatory filter feeding of small prey. Their name is believed to derive from Greek *mystakos (*moustache*)* and *ketos (*big fish*)*. They form a parvorder of Cetacea comprised of 4 extant families containing 14 different species.

Baleen Whales
14 species
4 families & 6 genera

Balaenidae
- Bowhead Whale
- Right Whales
  ...4 species

Balaenopteridae
- Blue Whale
- Fin Whale
- Sei Whale
- Minke Whale
- Humpback Whale
  ...9 species

Cetotheriidae
- Pygmy Right Whale

Eschrichtiidae
- Gray Whale

The waters around the Azores are home to six species of baleen whales, that occur during different times of the year depending on feeding and migration patterns. Mysticetes tend to stay at shallower depths than odontocetes. They are usually very large (with the females growing larger than the males), have two blowholes, and symmetrical skulls. They do not have any organ responsible for echolocation (melon), however they have very developed acoustic vocalizations. They are mostly found in smaller pods and most make sizeable annual migrations.

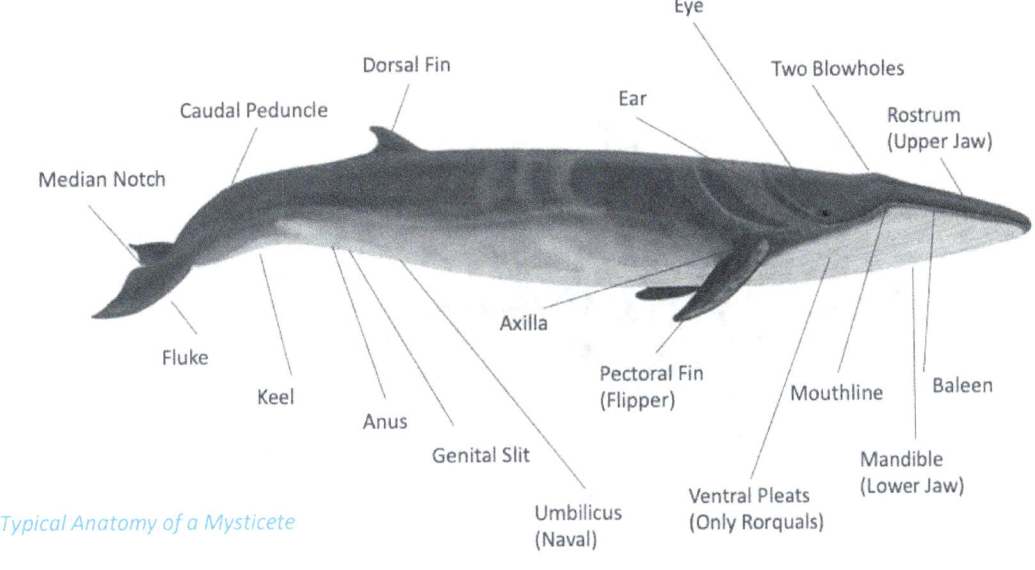

Typical Anatomy of a Mysticete

The Whale Research Assistant Program, 1st Edition

Odontocetes

Odontocetes are the toothed whales, feeding largely on fish, squid, and in some cases other mammals. Their name derives from the Greek *odontos (*tooth) and *ketos (*big fish*)*. They form a parvorder of Cetacea which includes whales, dolphins and porpoises. There are currently 76 recognized species, spanning 10 extant families.

Toothed Whales
76 species
10 families & 34 genera

Kogiidae
- Pygmy Sperm Whales
- Dwarf Sperm Whales

Physeteridae
- Sperm Whale

Delphinidae
- Oceanic Dolphins
  40 species
  - Orca
  - Pilot Whale
  - Bottlenose Dolphin
  - Common Dolphin
  - Spotted Dolphin
  - …etc.

Ziphiidae
- Beaked Whales
  …22 species

Monodontidae
- Beluga
- Narwhal

Phocoenidae
- Porpoises
  …7 species

Platanistoidea
- River Dolphins
  …5 species

There are regular sightings of 9 species of odontocetes in the Azores. Many occur as residents whilst others are transient populations. They are often smaller than mysticetes (with the exception of the sperm whale) with variable sexual dimorphism. Odontocetes tend to live in larger groups with more complex social organization than mysticetes. They chase and capture their food and dive much deeper than mysticetes. They have one crescent-shaped blowhole and a symmetrical skull. At the front of their head they have a melon: a bulging fatty area used for echolocation.

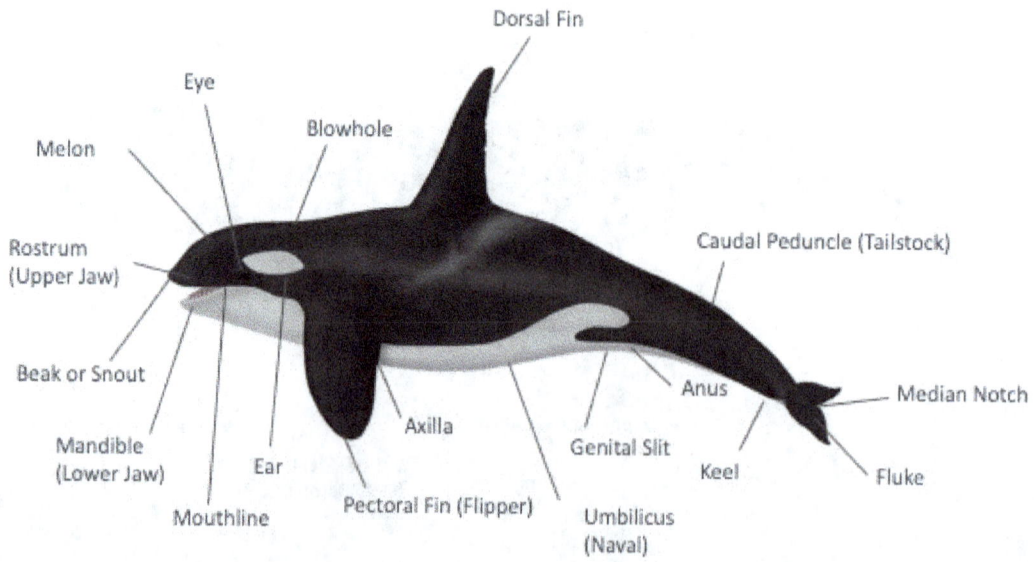

Typical Anatomy of an Odontocete

On occasion there are sightings of the Ziphiidae family of odontocetes, known as beaked whales. Little is known about this family and sightings are rare and unobtrusive, since they spend most of their time far offshore, at depths. Their bodies are spindle-shaped, wider in the middle and tapering at each end, with beaks of varying sizes and shapes. The dorsal fin is small; flippers are short and tucked into 'flipper pockets', to help reduce drag. The teeth of most females do not erupt and in males only one of two pairs of teeth erupt in the lower jaw, with none on the upper. They therefore feed by sucking their prey. When sighted in the Azores the following are most commonly noted:

Cuvier's beaked whale . *Ziphius cavirostris*

Sowerby's beaked whale . *Mesoplodon bidens*

Gervais' beaked whale . *Mesoplodon europaeus*

Northern Bottlenose Whale . *Hyperoodon ampullatus*

Cetacean Occurrence in the Azores

The Azores is the most remote archipelago in the North Atlantic, consisting of 9 volcanic islands and numerous seamounts. It sits on the mid-Atlantic ridge at a junction of three tectonic plates. It is a world renowned 'hot-spot' for whale observations, hosting some of the highest cetacean biodiversity in the world, for both resident and migratory species. The chances of sighting at least one species of cetacean whilst on an excursion is about 98%, whatever the time of year.

There are a number of physical, cyclical and seasonal phenomena that make the area around the islands a prime location for watching and researching these animals. The southeastern branch of the Gulf Stream, known as the Azores Current, and its eddies, flow south of the region- making the otherwise temperate area sub-tropical in part: creating a unique climate and oceanographic conditions that attract marine mammals.

Spring Phytoplankton Blooms

In the summer, the majority of mysticetes feed in colder waters at higher latitudes that are richer in planktonic organisms. In the winter they migrate to more temperate waters to mate and give birth. It is believed that the calm and warm tropical waters give calves a better chance of survival and are also areas with a lower risk of orca predation. En-route to these mating grounds, these whales must supplement their food intake with plankton encountered on their migration.

The Azores Archipelago is one of the waypoints used in their migration, where whales are seen feeding, suggesting that the region is tactically sought out in order to help these whales replenish energy.

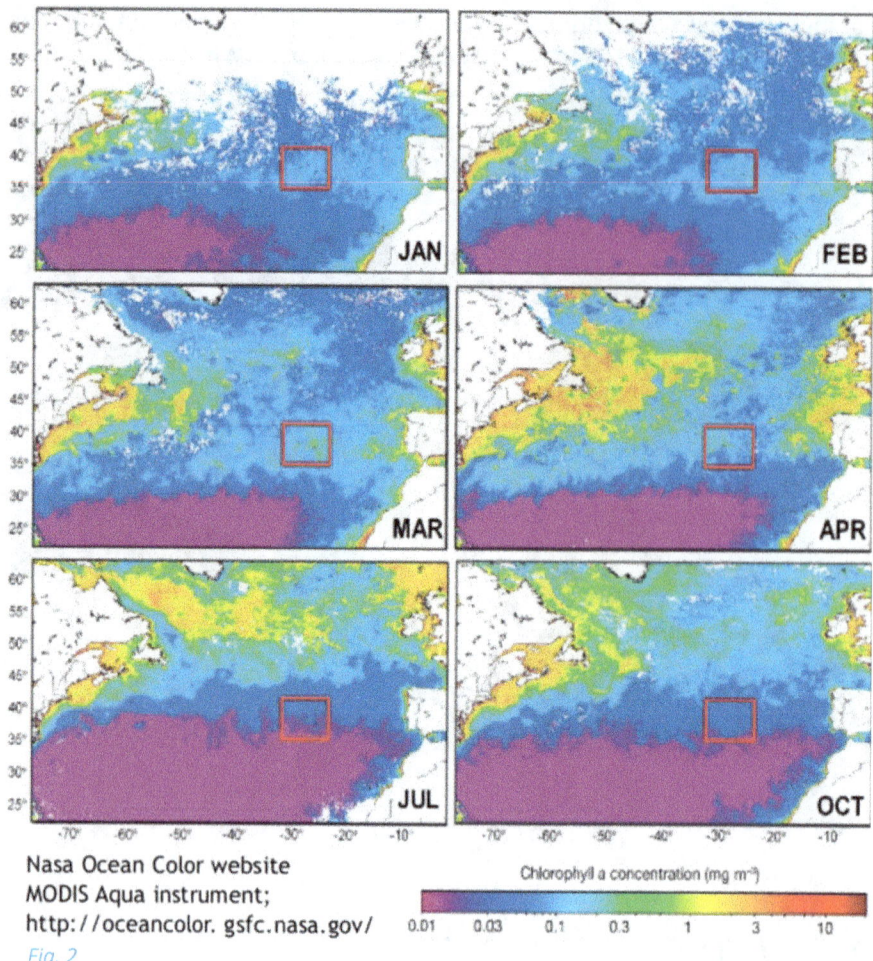

Nasa Ocean Color website
MODIS Aqua instrument;
http://oceancolor.gsfc.nasa.gov/

Fig. 2

The most observations of baleen whales are in the spring, when waters are highly productive due to phytoplankton blooms (Fig. 3). Since phytoplankton, just like terrestrial plants, use chlorophyll to photosynthesize, we can measure the concentration of chlorophyll-a in the water as an indicator of the presence of phytoplankton (Fig. 2). Growth rates are in conjunction with nutrient supply and light availability. When these two parts are optimized, rapid growth in phytoplankton can occur, leading to large biomass accumulations in the upper water column: a phenomenon referred to as a bloom. This can be important for global carbon budgets, but also drives bottom-up effects on the pelagic food chain.

Fig. 3

Colder waters tend to have more nutrients than warm waters, so phytoplankton tend to be more plentiful where the waters are cold. When surface waters are cold, it is easier for deeper water to rise, bringing nutrients to the photic zone through the process of upwelling. Throughout winter this process can nutrify the photic zones around the Azores, satisfying upper water layers.

However, especially in the polar regions- when there is less sunlight- phytoplankton is unable to grow as much. Once surface light increases in the spring, phytoplankton will begin to flourish and bloom in higher concentrations. The bloom ends once nutrients are depleted by phytoplankton (bottom-up control) and their subsequent consumption by predators.

Physical, Oligotrophic and Hydrographic Features

Across the vast Atlantic, much of the area is deep and abyssal. The Azores Archipelago has unique topographic features, as the islands rise from the abyssal plane at depths of around 2000 meters (1.25 miles). Diverse landforms are created around them, with more than 450 seamounts and other volcanic features- with complex oceanic circulation that supports a larger number of marine predators (Fig. 4). This undulating sea floor creates habitats for a number of deep-water species, specifically the giant squid: the main food source for sperm whales. The rising slopes of the sea floor at coastal areas around the Azores push up iron-rich and nutritious waters from the lower layers of the ocean, enhancing primary productivity. The seamounts can attract large aggregations of many marine species, which allows for numerous cetaceans to stay and feed year-round.

The spring enrichment has its origins in the upward movements of the Gulf Stream, its meanders and filaments, and the interaction of these oceanographic features with the islands. The sea surface chlorophyll follows an inverse pattern to that of sea surface temperatures. Nutrient-rich waters are often colder when they reach the surface, compared with their surroundings, hence the spring peak in phytoplankton coincides with seasonal lows of surface temperatures.

The North Atlantic is considered to be one of the most nutrient rich areas in the world. However, the Azores lies in a region with oligotrophic waters, with areas in the northern sector being more productive than those in the south. Typical of open-ocean oligotrophic sub-tropical regions, where light is not a limiting factor, most nutrients are expected to surface via a dynamic process, such as a filament or eddy, which raises the nutrients to the euphotic zone.

Biological and nutrient impacts are also linked to the topographic disturbance of ocean flows by the islands known as 'island stirring' where they generate their own wake and thus create eddies. In the Azores archipelago, where island groupings are close together and separated by deep water channels, the individual island effect (wake) is constricted, enhancing the phenomenon.

The biological enrichment is seasonal and dependent on the shallowing of the nutricline; it often reaches the islands from the north of the Gulf Stream, or from the east (Azores Current). This far-field enrichment strongly surpasses the intensity of the observed local enrichment, with higher levels observed north of the islands. The central group of islands has the largest oceanic imprint both in size and number of islands, therefore has a higher capacity to stir the incoming flow, inducing a shallowing of the nutricline and/or "capturing" the biologically enriched incoming system of eddies and fronts.

Upwelling is often intensified at coastlines due to the divergence of water. As winds diverge water, it creates a space and acts as suction on deep nutrient-rich waters in order to fill the space, known as Ekman suction. This process promotes phytoplankton blooms. Ekman suction occurs on the coastline of the Azores and in the open ocean surrounding it. In the same way, downwelling can occur as warm, nutrient-poor surface waters are pushed downwards, in a process called Ekman pumping.

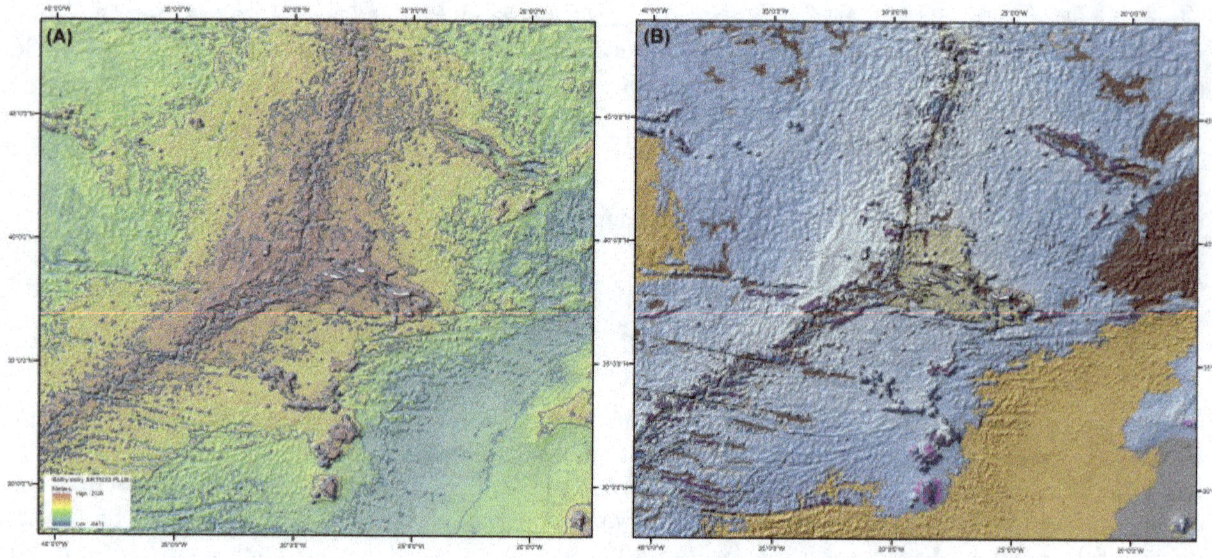

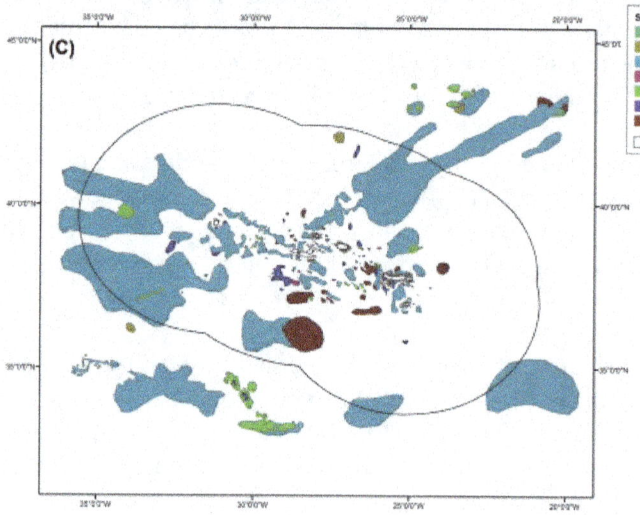

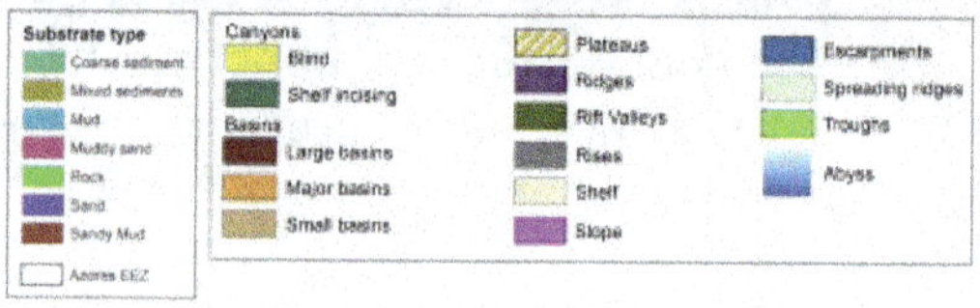

(A) Bathymetry of the study area for which all the derivatives were based on; (B) geomorphologic structures of the seafloor based on Harris et al. (2014); (C) seabed substrate type

Fig. 4

3D bathymetric map of the oriental group of the Azores.

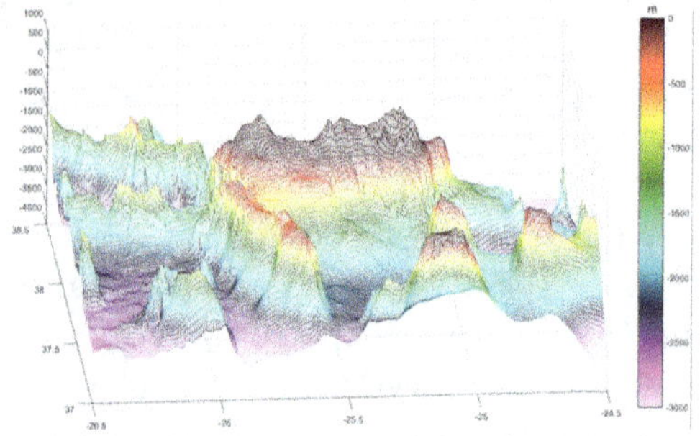

González García L, Pierce GJ, Autret E, Torres-Palenzuela JM (2018) Multi-scale habitat preference analyses for Azorean blue whales

Navigational Reference

The topography of the Azores also plays an important role for many migrating species of mostly mysticetes. The unique sequence of sea mounts around the Mid-Atlantic Ridge make it a key navigational reference for many cetaceans who travel between feeding and breeding grounds.

Tagging of such cetaceans has shown that the islands of the Azores often serve as a key reference point in these migrations, demonstrated by seasonal sightings of blue, fin, humpback and sei whales, to name a few. Changes in the composition of the sea floor and the rising of mounts, from otherwise featureless blue water, can prove vital to giving directional assistance.

The islands of the Azores, as well as the 450 other seamounts and volcanic features, rise from the abyssal depths of more than 2,000m (1.25 miles)

The Value of Cetaceans

As oceanic mammals, the value and benefits of whales is often not apparent. However, they are the unsung heroes of our ecosystems and are revered by many cultures and societies. Whales play a key role in atmospheric regulation and maintaining productivity of multiple trophic levels: enhancing primary productivity as well as keeping balance, as both top predators and prey, in the upper levels of the food chain.

The Whale Pump

Whales have often been coined as 'ecosystem engineers'. They have been shown to have a large role in enhancing the primary productivity of marine ecosystems by concentrating nutrients near the surface.

Deep waters become rich in nutrients through showers of marine snow that form a downward flux of organic matter and detritus, falling from upper layers of waters to the deep ocean. Zooplankton also produce sinking fecal particles at depths during migration cycles, adding nitrogen, phosphorus and iron to deep waters. Primary production can be limited by the availability of heavy metals such as iron, which is pulled down by gravitational forces to the ocean abyss. This process is known as a downward pump of nutrients.

Whales feed at depths, but defecate at the surface, releasing fecal plume nutrients in the photic zone of the water column. Also, within this movement, whales push nutrients from the bottom of the ocean to the surface, through a process known as upwelling. Whales do not absorb much of their dietary iron and so their fecal plumes are highly concentrated with this limiting nutrient. These processes act as an upward biological pump, known as the 'Whale Pump'. The pump helps mix the water column and spread essential nutrients through the marine layers.

The cycle brings nutrients back to the photic zone where it fertilizes phytoplankton, the base of the food chain. Phytoplankton needs to be near the surface in order to access the sun's energy for photosynthesis. The nutrients brought up through the whale pump and whale fecal plumes stimulate growth, causing the phytoplankton to bloom.

The phytoplankton absorbs carbon from the atmosphere which enhances the carbon capture process. As a byproduct, the phytoplankton release oxygen in collectively huge quantities, with estimates suggesting around half of the global oxygen supplies are sourced here. Therefore, by enhancing this cycle, whales are

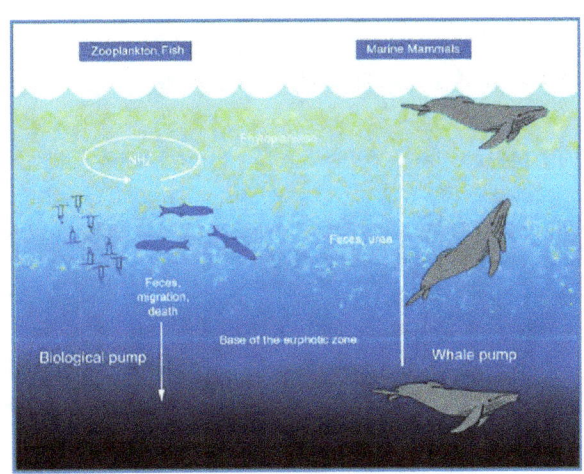

Roman and McCarthy, 2010

crucial to oxygen production, ocean carbon absorption and climate change stabilization.

The phytoplankton form the foundation of the trophic structure, providing a food source for many small marine animals such as krill and marine flora. Blooms of phytoplankton, stimulated by the whale pump, create a positive feedback loop within a healthy marine ecosystem. By creating these areas of high primary productivity where whales aggregate to feed, mate and give birth, secondary productivity is stimulated, creating greater abundancies of fish. Whales can also horizontally transport nutrients thousands of miles from productive feeding areas at high latitudes to less productive calving areas at lower latitudes, expanding the areas of high primary and secondary productivity across vast oceans.

Whale Fall

When a whale dies, its huge carcass can fall to the bathyal or abyssal zones, which begins to provide a bounty of nutrients to the otherwise barren, deep water ecosystems on the seafloor. The colder temperatures and high pressures slow decomposition rates and increase gas solubility, allowing the carcass to remain intact. The fall provides a sudden concentration of food and sustenance to deep-sea organisms for decades to come.

The carcass is devoured in different

Ocean Exploration Trust and NOAA ONMS

stages. In the initial phase scavengers, such as sharks and hagfish, detect the scent and swim from afar to consume the soft tissue, whilst detritus enriches the sediments nearby. This phase can last up to two years, after which an enrichment opportunist phase begins, where worms, crustaceans and mollusks feed on the leftover blubber and burrow into the nutrient enriched sediment beneath the carcass. Finally, the sulphophilic stage begins and can last for several decades. At this point, with only the bones remaining, bacteria start to anaerobically break down the lipids trapped inside, generating sulfides from the decay of organic compounds. The bacterial mats formed provide nourishment to organisms such as mussels, limpets, sea snails and clams.

Note that, should the carcass fall in shallower waters, then it will be consumed by a high incidence of scavenger hunters over a relatively short period of time and the warmer waters will hasten the decomposition process. In these instances, although still beneficial to many organisms, we do not see the full 'whale fall' process as described above.

Human Cultural Importance

Whales play large roles in many local folklores and are believed by some to guide travelers across the ocean. Over the course of history whales have been viewed as legends, gods and deities in many cultures and mythologies.

In Hawaiian a traditional chant, the "Hanau ka palaoa noho i kai", teaches that whales are part of both darkness and light; divine and physical. The coastlines on which whales were washed up or stranded were held especially sacred and thus were protected by the chiefs and priest. The whale is believed to be the animal form of the Hawaiian ocean god Kanaloa.

For Maori tribes, whales are descendants of the god of the

oceans, Tangaroa, and were deemed to be sacred, supernatural beings. For migrating tribes, whales were seen as a sign, indicating that they should settle in a particular place. In Australia, the aboriginal people recorded grand stories of whales in rock engravings and totems.

In Native American culture, whales, especially the orca, are known as the 'Lords of the Oceans', symbolizing family, romance, longevity, harmony, travel, community and protection. They are said to protect those who travel away from home and lead them back when the time comes.

Economic & Social Importance

Eco-tourism has become a vital part of coastal economies and is growing at a rate of 1.3%. Whale watching is a huge part of ocean tourism, with great economic importance to many locations around the world. People are fascinated by these oceanic mammals and intrigued by their intelligence. Billions are spent by people worldwide attempting to see the largest animals on earth in their natural habitats.

In 2012, the whale watching industry generated approximately $2 billion (US) in revenue and supported roughly 13,000 jobs worldwide. In order to help conserve the oceans, we often need 'poster animals' that can infer an emotional response in humans. People often feel a strong emotional

connection to whales and dolphins. Conservationists can use this sentiment to help increase stewardship and action to not only save these animals, but the ocean as a whole. Many people leave the Azores after a whale observation trip with a greater appreciation of the ocean and understand the need to protect it.

Dive Azores, Researchers

Predators and Prey

Whales are both primary consumers and apex hunters and therefore play a key role in providing top-down and bottom-up controls within the trophic structure. Depending on species, the types of foods a whale consumes can vary from small krill and plankton to large seals and giant squid.

Orcas are sometimes found hunting smaller cetaceans and whale calves and, in rare instances, predate on larger sharks as well. Occasionally pods of orca are seen attacking adult whales, however, are rarely successful. After many years of commercial whaling, the orca's diet is believed to have expanded due to dwindling numbers of whale calves, which has caused detrimental knock-on effects to other food webs. One example of this is the decrease in sea otter populations due to predation by whales. This has in turn caused an increase in the abundance of herbivorous sea urchins- the otters' prey- that are now destroying kelp forests: an example of how important the population dynamics of whales can be to the overall health of the complex marine ecosystem.

Baleen whales filter feed on phytoplankton and smaller ocean organisms in massive quantities. This forms a key role in carbon cycling, which is sequestered out of the atmosphere and naturally accumulates in whales' bodies.

Baleen whale feeding

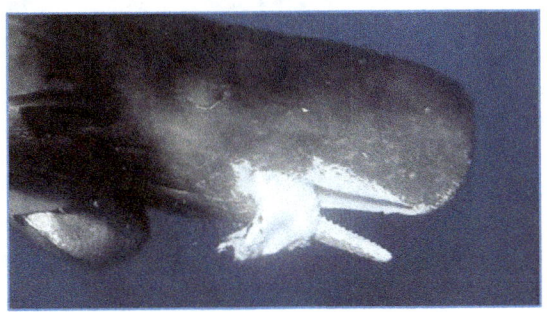
Toothed whale feeding

A single adult blue whale is capable of eating around 4 tonnes of krill per day, requiring around 1.5 million calories: more than any other animal

Communication

The study of communication amongst whales is an extremely challenging science, hence little is known for the vast majority of species. The difficulties come with regards to locating the animals and also understanding the intended recipient of the vocalizations, who may be many miles away. Non-invasive tagging technology is limited too, if not combined with observational records. Another consideration is simply the influence of the observer on any animal behavior, especially for cetaceans, when noisy boats are involved. There are, however, some things which science has discovered.

Cetaceans communicate with one another in three different ways:

Visual Communication

Visual communication comes with its limits. The turbidity of the water, light penetration, and availability of light at depth can limit the ways in which visual communication can be made over distance, often limited to just tens of meters in the best conditions.

Visual displays are produced through various behaviors, including simple expressions of sexual dimorphic features, body postures and coloration patterns- to more complex sequencing to indicate reproductive intent, situational context, age, etc.

Coloration patterns can appear by uniformity and express visual cues. Countershading is a process of camouflage used especially by small cetaceans to evade predators and avoid being seen by prey. Countershading is when an animal is dark on top to blend in with the seabed or deep ocean, and light on the bottom to blend in with sun glints or shallow waters. Movement can also be a form of communication for individual species: in the case of swimming speed, grouping behaviors, and acrobatic displays at the surface. Visual displays of sexual dimorphic features can play an important role in regulating social signaling and mating. This is best represented in the males of several beaked whales, where the protrusion of the lower teeth outside of the mouth may be the only visual indicator of sex.

Bottlenose dolphins playing

Dolphins open their mouths and jaw clap as a threat display.

Gestures and non-vocal sounds are also important in cetaceans, such as open-jaw clap threat displays; breaches, flared pectoral fins, tail lobs, tail slaps, etc. Posture and behaviors may be giving signs that there are predators or prey in the areas, in order to synchronise actions amongst individuals, to coordinate the group, or simply for social interaction. The 'S-posture' has been a frequently reported posture, where a cetacean abruptly stops forward motion; flexes the fluke and sustains an arched 's' shape- generally used in courtship or to display aggression.

Tactile Sensing

Tactile communication is also known as touch and can include rubbing of the snout, rostrum, flippers, fins and even the whole body during social interactions. More aggressive tactile signals include ramming, raking, butting and biting.

On the skin of mysticetes, mostly around the mouth, there are patches of hairs called a *vibrissae* that form a somatosensory system. They function as tactile sensors, giving the whale a well-developed sense of touch. Odontocetes lose these vibrissae during foetus development, leaving behind just the pores or vibrissal crypts in some species.

Large scarring on Risso's dolphins

Close pod of orcas hunting

Acoustics

Through tactile sensing and vision, cetaceans are only able to communicate in close proximity to one another. However, through using acoustics they are able to communicate over vast distances- tens or even hundreds of kilometers apart. Acoustics play a large role in the lives of cetaceans, since sound travels underwater more efficiently than any other form of energy. The limitation of vision to only tens of meters in the best conditions- due to turbidity and by light at depths and at night- means sounds are used for countless reasons: not only in communication but in mate selection, navigation, hunting and group cohesion.

Sperm whales at the surface

All cetaceans have sensitive hearing and a great part of their communication is achieved acoustically by producing different types of sounds: clicks, whistles, moans and songs. Cetaceans show a great amount of diversity in their communication patterns, but research to date has mainly focused on just a few species.

The mysticetes, unlike the odontocetes, tend to undergo annual migrations. Analysis of communication is notoriously difficult, since the spatial scale that must be included to study and observe both the senders and receivers of acoustic communications is large. Saying this, however, one of the most popular studies of whale communications has been in the acoustic 'songs' produced by several baleen whales, especially the humpback (*Megaptera novaeangliae*). These patterned vocal sequences of different complexity have yet to be observed in many odontocete species.

Humpbacks often breed in accessible areas near shore and became an attractive area of study due to their production of 'whale song', a frequency of acoustics well within range of human hearing. The songs are only sung by males and follow a structure, just like that of a sheet of music. They consist of units which- combined together- form phases, producing themes in repeating patterns that can last over 30 minutes. After several decades of analysis, there is still no consensus regarding the functionality of humpback whale song, other than a mutual agreement that it is likely a form

Researchers administering tags

of sexual display, since the majority of recordings have been performed by lone males at breeding grounds. However, males have also been recorded singing on migration and in the company of females.

An intriguing study of humpback whale song is its ability to change over time, alterations being clearly derived from vocal learning. Song derivations and types have been shown to pass on through populations and 'top the charts' in different areas as they spread across the oceans. This way of learning or culture was once thought to be innately human. It has since been proven therefore that whales in fact have culture also, which they pass on socially and cognitively.

The larger mysticetes, the blue and fin whales, can produce sounds loud enough and at a low enough frequency (< 20 Hz) that they could be theoretically heard across ocean basins. Data suggests however that most calls range around 90km (56 miles), so that migrating animals can be reached by calls in a biologically relevant time frame. Blue whale songs appear to be associated with feeding and remain fairly stable over decades, therefore don't exhibit the same vocal learning as with humpbacks.

Where songs play a role in reproduction, any group or individual can use specific signals within social relationships in order to maintain bonds: a characteristic much more associated to odontocetes. These whistles and clicks (also used in echolocation) can express both joy and warnings. Some beaked whales produce acoustics only at depth, which has also been suggested as an anti-predator adaptation.

Diving Physiology

Many species of whales are able to dive to extreme depths, withstanding the extreme pressures of the abyss. The deepest known diver is the Cuvier's beaked whale, which can reach limits of 2,992 m (9,816 ft) on a single breath, remaining underwater for 138 minutes. Sperm whales routinely dive between 500–1000 m (1600–3300 ft) and, like other deep divers, are doing so to forage for food. For each 100 m (330 ft) of depth, animals experience pressures 10 times that on the surface. Pressure increases by 1 atmosphere for every 10 meters of ocean depth, meaning that animals diving to 1000m experience 100 times the pressure at the surface.

Cetaceans have extremely efficient lungs, using up to 90% of their lung volume in a single breath, compared to humans, who only use 15-20%. To avoid the complete compression of the air in their lungs when deep diving, they have specially adapted rib cages that fold to collapse their lungs and reduce air pockets.

Cetaceans must also slow down their heart rates significantly and restrict muscle movements by simply freefalling. Prior to the dive, they exhale 90% of the air in their lungs to reduce buoyancy, making it easier to dive. However, doing so reduces the total amount available oxygen in the lungs. So, whales conserve oxygen by shutting off blood flow to their extremities and organs not necessary for diving (i.e. liver, kidneys, and intestines) and redirect it to core functions (i.e. brain, heart, muscles). They achieve this by well developed reservoirs of oxygenated blood called the *rete mirabile*. These provide bypasses that enable cetaceans to isolate skeletal muscle circulation, while using the oxygen stored in the remaining blood to maintain heart and brain functionality.

At depth, cetaceans need oxygen to maneuver and hunt. To facilitate this, they are adapted to store high levels of oxygen in their blood and have a high blood-to-body volume ratio. They are also able to store more oxygen in their muscles than other mammals, since the levels of myoglobin (the red protein that carries and stores oxygen) in the muscles is ten times more concentrated. The myoglobin in these cetaceans is also positively charged so that large amounts can be stored without clumping together, which could cause serious diseases seen in humans- such as diabetes and Alzheimer's.

It is still unknown how deep diving cetaceans deal with a disorder known as decompression sickness (DCS) caused by the oversaturation of dissolved nitrogen at high pressures forming bubbles when pressure decreases upon a fast ascent. A number of autopsies of dead whales have shown signs of decompression damage, confirming that it is a condition that must be mitigated within the dive. Although susceptible to the condition, it is assumed that deep diving whales have evolved to ascend slowly. Disruption or panic may interfere with the dive and deep foraging, sometimes causing cetaceans to ascend faster than usual and suffer the effects of DCS. This could sometimes be the cause of death in cetaceans that wash up with no visible signs of illness or injury. Although still not fully understood, we know that since nitrogen is still contained in the breath hold and is pressurized at depth, accumulation can still occur in the blood, which could result in DCS.

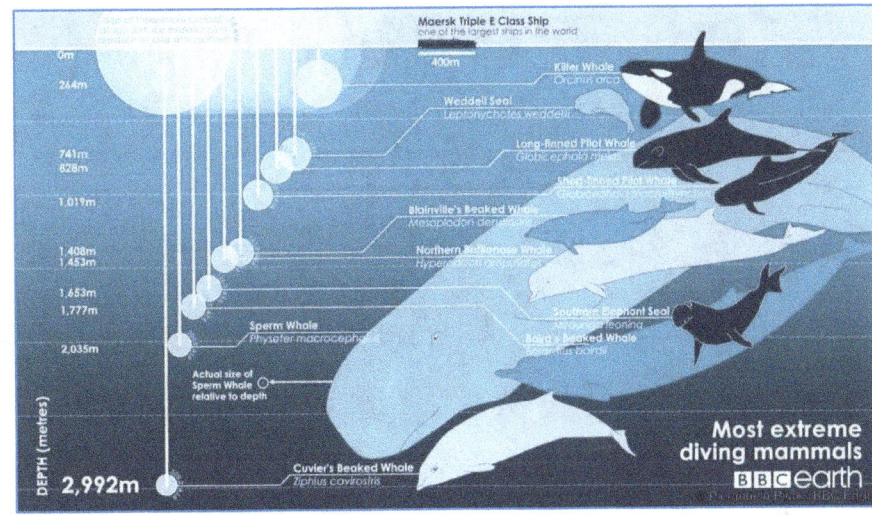

Feeding

Odontocetes

Toothed whales' teeth are shaped to grab and disable prey, but not to chew. Their teeth are mostly identical, with varying numbers across species. For some species, and sexes within species, the teeth do not ever erupt. If prey is too large to be swallowed intact, it is ripped into chunks by gripping and shaking.

Odontocetes are considered to be 'hunters': seeking out prey, chasing them in cooperative group hunts, with hopes to capture and consume. The targeted prey species varies greatly within the odontocetes, consuming everything from numerous types of fish, shark, cephalopods and crustaceans, to other mammals like seals and penguins. The orca is also known to predate upon other cetaceans, especially young calves.

It is believed that most odontocetes rely on the use of echolocation, a highly sophisticated sonar system, to navigate and detect their prey. Ultrasounds, high frequency click signals, are created by squeezing air through the nasal passages, that are then passed into the melon at the front of the odontocete's head, which focuses the sound waves into a beam. If these waves hit something, they bounce or 'echo' off the object which is rebounded back to the whale's lower jaw, which contains special sound conducting tissue pads connected to the inner ear, along a fine channel of fat. The echo is absorbed and processed into a 'sound picture' that displays the shape, size, density and direction of movement of the prey. This specially developed sense allows odontocetes to hunt in waters of high turbidity and at depths where no light penetrates. Some are able to stun their prey prior to the bite using high-energy bursts of 'sound'.

Sperm whale spy-hopping

Orca pack hunting seals off the ice

Feeding

Mysticetes

In order to locate prey, mysticetes use low frequency sounds to navigate ocean basins (biomagnetic orientation) and communicate with conspecifics. Instead of teeth, mysticetes have baleen: comb-like keratin plates hanging down from the upper jaw. When feeding, mysticetes are known to target and herd prey into large balls or schools, so they are able to engulf massive quantities at once. Feeding in bulk on dense aggregations of prey is required to support their huge body masses. When prey is detected, mysticetes open their mouths, taking in huge quantities of water. Upon closing, they keep their mouths partially open to push out the water so that it strains through the baleen along both sides, separating out the prey to swallow. Feeding in this way, mysticetes utilize the following strategies:

- Gulping
    - Engulfing
    - Lunging
    - Skimming
- Bottom Feeding

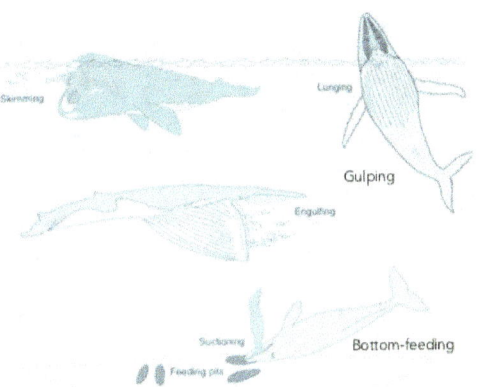

Lunging and engulfing is exclusively performed by Rorquals (cetaceans of the genus *Balaenopteridae*), who have ventral grooves whose pleats are able to stretch to enlarge the mouth. When rorquals accelerate towards a dense area of prey, they open their mouths wide, creating a large gape angle (up to 90°). Upon opening the mouth, the pressure generated from the extra drag forces water inside, expanding the oral cavity to several times its normal size. Within 2-3 seconds, at 5 knots, a humpback whale can engulf 60 tonnes (66 tons) of water: more than its own body weight. This makes the whale look very bloated but, within a minute, the water is filtered out, leaving the prey inside to be swallowed. Lunging is performed vertically upwards, targeting a density of prey near the surface, whereas engulfing occurs more horizontally within the water column.

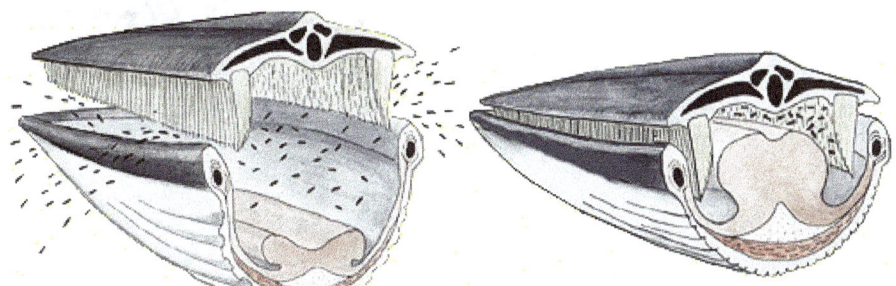

The rorquals are able to employ this feeding strategy because of number of anatomical features and specialized jaw joints within a kinetic skull. They have large, bilaterally separated, mandibles (lower jaws) which are connected to the base of the skull by a dense matrix of elastic fibres and oil infused cartilage, which allows for a large gape angle. They are able to rotate their jaws outward due to another flexible joint at the front of the jaw where the mandibles meet. A flexible and flaccid tongue is able to easily invert itself, when water rushes in, and then retreat through the floor of the mouth to help form the large oral cavity, aid in the straining process, and stop water from being swallowed. The ventral grooves themselves have strong muscles with a sensory system of mechanoreceptors to control incoming flow of water and manage the engulfment actions.

Each time a cetacean uses a gulping feeding strategy, it expends a large amount of energy, both in accelerating and in the drag created, which brings momentum to a near halt: requiring more energy for any reaccelerating for subsequent gulps. Dives for engulfing can only last around 15 minutes before oxygen stores are depleted and are followed by long surface intervals. Each lunge or engulfment must therefore be carefully calculated to ensure that a cetacean targets dense enough aggregations of prey to make positive energy gains. A blue whale can take in half a million calories in just one mouthful. To this end, whales have been documented swimming through masses of prey that are too small without opening their mouths to feed.

Bubble-net feeding of humpback whales

Baleen whale gulp feeding

The non-rorqual mysticetes, such as bowhead and right whales (as well as the sei whale) use the other strategy: skimming. Once they have located a concentration of prey, they slowly swim through it with their mouths open, allowing water to enter through a gap in the baleen at the front and filter out through the two baleen racks. Since the success of this strategy is dependent on the surface area of the baleen, the baleen plates are around twice the size of those who gulp feed, reaching around 2m (6.5ft). Skimming occurs at the surface and at depths, but requires slow speeds to ensure efficient use of oxygen stores.

The other method of feeding, known as bottom feeding, has only been recorded in gray whales. It is performed in shallow waters by turning sideways on the sea floor, digging the rostrum into the sediment, then creating a pulsing suction using a muscular and maneuverable tongue. As it rises upwards, water and sediment is strained out through short, thick baleen plates with frayed inner edges, isolating prey items in the mouth that are then swallowed. The gray whale searches for small bottom-dwelling crustaceans (amphipods, isopods, mysids), worms and mollusks. The process often creates plumes, visible from the surface, and grooved feeding tracks on the ocean floor.

> Research has shown that gray whales are 'right-handed', since the baleen on the right side shows more wear, and the head has fewer barnacles and skin abrasions.

Gray whale bottom feeding

Identifying Cetaceans

Identifying cetaceans can be a challenging task, even for the most experienced cetologists. There are a number of key attributes that we use to help in the identification of both species and the recording of individuals. Even with this toolkit though, on occasion, some sightings result in an unidentified recording. Knowledge of the area you are observing, the season you are in, and topographical maps can help you calculate the likelihood of specific sightings, however you must always be prepared for any encounter. No whale observation excursion is predictable, and every cetacean has a personality and will of its own.

When monitoring cetaceans, you should aim to gather as much information and data on all of the following features, in order to make a scientifically viable recording, as well as getting the satisfaction of making a correct identification. Do not be disheartened on your initial monitoring excursions at sea if you are only able to make recordings such as 'large whale' or 'small group of white dolphins with lots of scratch marks'; just make as many notes as you can. Your instructor will be on hand to point out different features and will provide a debrief where you can look at photos to confirm species and hopefully gather enough details to record an individual. Over time your knowledge will grow, and observational skills will be honed so that identification will come much more easily to you.

Size

It is notoriously difficult to size animals in water, since your ability to get a close enough distance away for a sufficient amount of time is reduced, and any underwater photography will have the magnifying effects of light refraction applied. At sea when estimating the size of a cetacean we try to use reference points, such as the boat, as a comparable measure of the animal. We then place the animal into 4 categories of size:

- Small: **≤ 3 m** (≤10 ft)
- Medium: **>3 – ≤10 m** (>10 – ≤33 ft)
- Large: **>10 – ≤15 m** (>33 – ≤49 ft)
- Extra Large: **>15 m** (>49 ft)

Blue whale diving

Estimating size can also be problematic since only portions of the animal are visible at any one time, especially the case for the larger cetaceans. It is important therefore to gain the knowledge of size brackets for different species at their different life stages. Your instructor will also be on hand to help assist sizing cetaceans when at sea.

Unusual Features

Distinct anatomical features and markings of different species of cetacean can be used to quickly identify them. For example, we are able to quickly identify an orca by their black, straight, tall dorsal fin that can grow up to nearly 2 m (6 ft), towering above the surface of the water, with a distinct gray/white saddle behind it. Sperm whales are also quickly identified when you see their wrinkly, prune like skin.

Wrinkled skin of sperm whale

Dorsal Fin

The shape, position on the body, and size of the dorsal fin varies between different cetaceans, and can come with a variety of different markings, notches, nicks and colorings. The dorsal is often described first in one of four shapes:

- Falcate – curved backwards, sometimes referred to as sickle-shaped
- Triangular – in a straight edged scalene triangular shape
- Elongated – tall and pointed in an isosceles triangular shape
- Rounded – forming a rounded top and smooth sides

Some dorsal fin shapes can also be more curved or hooked back than others and some may fall in between falcate and triangular, but these can be important descriptive features to note. The dorsal fin may also be completely mutilated or deformed, marked, or undefinable due to the angle of the sighting or photograph.

When describing the dorsal fin, we refer to the front as the 'leading edge' and the back as the 'trailing edge' and where these edges reach the boundary with the body, the anterior and posterior insertion points, respectively. Most of the time markings like nicks and notches are found on the trailing edge of the dorsal.

Blow

In larger whales the blow and spray can be particularly distinctive, and is very useful for spotting cetaceans while surveying from land or sea. Spray varies in height and intensity- depending on the size of the cetacean; when the blow occurs during its surfacing (the first blow after a dive tending to have the most power); and the behavior it is exhibiting. External factors, such as wind speeds and rain can greatly alter the blow and light conditions, and boat positioning can greatly influence visibility of the plume.

Flukes

Flukes are especially important in the larger whales, where some species lift their flukes into the air or 'fluke up' as part of their pre-dive sequence. For species such as sperm and humpback whales, fluke observations are the main form of photo identification for individuals, due to distinct colors and markings acting like 'fingerprints'. The following details should be noted:

- Shape of the fluke tips – pointed, rounded, incomplete, mutilated etc.
- Shape of the trailing edges of the left and right fluke – undulations, notches, toothmarks, scallops etc.
- Shape of the leading edges of the left and right fluke – undulations, notches etc.
- Presence of a median notch between trailing edges of the left and right flukes
- Unique markings and color patterns

Body Shape

Similar to the issue of sizing cetaceans, it is rare to see enough of the animal in order to gather a full impression of their shape. Therefore, shape is often limited in use to describe the dorsal fin or tail fluke. However, shape can be useful when assessing whether a female is pregnant or the whether different animals are proportionally slimmer or stockier than others. The shape of the melon, mouthline and rostrum of the animal can also help as distinguishing features of different species.

Color and Marking

Color and markings are an important part of cetacean identification, as many species can be identified by their unusual colorings and patterns, as well as scarring from different activities. Colored patches, body strips and spots all form distinct features of different species. Unusual marks, scars, lesions, raking wounds and notches can also be vital data points when recording individuals for population studies.

Note that light can help or hinder you when recording color, and so is not usually a reliable identification point by itself. Water turbidity may also make it harder to see cetaceans that remain more submerged.

Scarred fluke of sperm whale (probably caused by boat collision)

White lower right jaw of fin whale (left side dark)

Diving Sequence

The way in which different species move and operate between breaking the surface from a dive and diving again can be quite distinctive. The following observations can be made:

- The angle at which the head breaks the surface
- What part and how much of the head is visible
- Whether the dorsal fin and blow hole are surfaced at the same time
- Breath intervals
- Number of shorter dives
- Number of breaths before a deep dive
- How arched, if at all, the back of the cetacean becomes when diving
- Whether the cetacean flukes up when deep diving
- Deep and shallow dive times

Group Size

Some species are often seen to be solitary or in small groups whilst others are highly gregarious, living in large groups. Just like estimating a cetacean's size, the number within a group can also be difficult, especially for larger groups where there are different proportions underwater/at the surface at any one time and when splinter groups are formed.

Beak

Mostly useful when trying to identify beaked whales where the prominence of a beak can be a useful indicator. River dolphins and a large proportion of oceanic dolphins also have prominent beaks and so observing the presence or absence of a beak can be important here when making a distinction. Beak length and whether there is a smooth transition or distinct crease at the border between the top of the head and the beak are features to take note of.

Behavior

When we talk about behavior, often we are talking about how active a species of cetacean is on the surface and its reaction or shyness towards observing boats. Any peculiar behaviors are also useful for identification and as an interest to science. We often observe specific behaviors of some cetaceans to be associated with socializing, feeding and mating, that are unique to or common for that species.

Cetacean Behavior

As mentioned previously as part of cetacean identification, behaviors are not only an important tool to positively identify a species or individual, but are useful and interesting observations for both whale enthusiasts and scientists. Cetaceans have diverse social networks and lifestyles that create visibly different behaviors. Some behaviors are exhibited whilst on the move, while others form part of stationary surface actions or performances. Apart from the respiration requirement, the development of certain surface behaviors have been established for communication, feeding, sexual displays and in some cases, just for fun. The majestic and energetic surface displays cetaceans exhibit are one of the main reasons humans flock to observe them, creating touristic interest as well as scientific.

Blowing

Blowing is the process of exhaling through the cetacean's blowhole upon surfacing from a dive, which explosively forms a cloud of mist. The size, dispersion and intensity of these blows varies amongst species. It can be quite distinct in some species, making it a valuable identification tool. In others it may be nondescript or not valuable for identification.

Breaching

Breaching is the act of leaping out of the water with the intent to clear it. It has been classified as any leap were a large proportion of the body (>40%) leaves the water. A lunging behavior is also seen where a smaller part of the cetacean's body leaves the water as an unintentional consequence of feeding.

Breaching is observed in many large whales, most notably in the humpback and sperm, who swim vertically from depth and head straight up out of the water. Another breaching method involves swimming close and parallel to the surface and, with a few strong tail strokes, propelling themselves out of the water. Upon landing, the cetacean usually turns on its side but sometimes 'belly flops'. Breaching is also seen in fin, blue, sei, minke, Bryde's, gray and right whales.

Breaching is energy intensive, particularly when it is done in a series, so it must be beneficial to the animal to have been selected for in evolution. The reason for breaching is unknown, however it is hypothesized to be a social behavior, or one to demonstrate physical fitness to exhibit dominance or sexual prowess. It is also possible that the large impact and sound upon entry is used to stun prey or remove parasites.

Breaching is often confused with porpoising from smaller cetaceans, who are easily able to completely clear the water.

Swimming Speeds

Cetaceans swim by vertically moving their tails using powerful epaxial and hypaxial muscles along the spine to drive their horizontal flukes up and down, propelling them through the water. Amongst cetaceans, most have a cruising speed of around 7 km/h (4 mph). The fastest cetacean is the sei whale which can reach surface speeds of 50 km/h (31 mph). Blue and fin whales can swim fast enough that observation boats must reach speeds of more than 30 km/h (18 mph) to reach them, whilst humpback, gray and right whales seldom swim faster than 9 km/h (5 mph) and sperm whales cruise around 7.5 km/h (5 mph) with spurts of up to 36 km/h (22 mph).

The speed at which cetaceans are moving tells you about their current behavior, be it hunting, socializing, evading predators, or simply conserving energy. Swim speed can also change when an observation vessel approaches, telling you a little about how the cetacean may be feeling or how shy they are.

Porpoising

This is a high-speed surface behavior where swimming near the water's edge is combined with long jumps; where jump length is roughly equal to the distance travelled when under the water. This is a behavior seen in smaller cetaceans such as dolphins and porpoises when travelling at speeds in excess of 16 km/h (10 mph). When at these high speeds, a lot of energy is exerted and so being in close vicinity to the surface provides greater ease to maintain respiration. The longer jumps also aid deeper breaths to ensure adequate oxygen supply.

Porpoising is exhibited when animals are involved in an important pursuit or as an escape tactic. Leaping out of the water is more energy efficient in retaining high speeds, due to reduced friction in air. However, some species such as spinner dolphins, rotate their bodies during porpoising, lending support to the theory it is not just solely a tactic employed to conserve energy, but may be part of communication or play.

Fluking

Just before a deep dive, some species of whales lift their flukes up out of the water in order to descend steeply rather than progressively, and so this is usually a behavior exhibited by those species who feed at greater depths. There are two types of fluking exhibited:

- **Fluke-Up:** Where the underside of the fluke is shown as it is lifted up high in the air (e.g. sperm and humpback whales)
- **Fluke-Down:** Where the fluke clears the water but remains turned down, not revealing the underside (e.g. blue whales)

Logging

As in the name, logging describes a behavior exhibited by whales, in particular right, sperm, pilot and humpback whales (especially when with calves) where they rest on the surface and appear like 'logs'. During this time the animal is floating horizontally, without moving, with the dorsal fin or parts of the back above the surface to aid breathing.

Whales can rest beneath the surface. However, they are required to perform unihemispheric slow-wave sleep: they retain conscious breathing by only resting one-half of their brain at a time. In these situations, they rise intermittently to the surface, breathe and exhibit logging behavior. Resting under the water is mainly done in a horizontal position, although sperm whales famously often choose to rest vertically.

Spyhopping

Spyhopping is a behavior in which a cetacean rises vertically upwards out of the water exposing its entire rostrum and head, and then holds this position. The whole movement is slow and controlled. Spyhopping is exhibited when a cetacean is curious about something, therefore its eyes will be slightly above or below the surface, holding themselves in this position for a minute or so at a time. The action is performed by positioning the pectoral fins and demonstrating acute buoyancy control.

The behavior is often seen in orca as they patrol ice floes, searching for prey, and when a cetacean becomes inquisitive about boats in the water around it. However, their eyes do not always break the surface, and in these instances, it is believed the whale may be trying in enhance its ability to hear sounds above the water.

Lobtailing

Unlike fluking, lobtailing is where a cetacean lifts its fluke out of the water only to slap it against the surface hard and fast, often several times. The sound from the slap can be heard for hundreds of meters underwater, hence is believed to be a form of non-vocal communication as well as a visual display. It is also thought that, amongst humpbacks, lobtailing may be used for foraging by scaring fish into tighter schools, making it easier to consume more in a single gulp.

Lobtailing is most commonly seen by sperm, humpback, right and gray whales. They position themselves facing downwards and slap the surface by bending their tail stocks. Dolphin species that also lobtail tend to make the slap by jerking their whole body either on their front or back.

Flippering/Flipper Slapping

Flippering is similar to lobtailing, where species of large whales use their pectoral fins to slap the water, for the same desired effects.

Bow-riding

Bow-riding occurs when cetaceans position themselves in the pressure wave at the bow or front of the boat. At high speeds cetaceans, specifically smaller odontocetes, will seek out the pressure waves and energy in order to body surf and essentially enjoy the free ride in the bow wave created by the vessel. Whilst surfing the wave they hold their flukes in a fixed plane, reducing the energy requirement of swimming.

Wake-riding

Wake-riding describes a behavior in which cetaceans jump and ride in the waves or wake behind boats. Cruising in the wakes reduces the energy requirements when swimming at high speeds. The strong kinetic energy created by the boat wake can propel the cetacean along.

Playing

A lot of the behaviors we have described above have been attributed to survival purposes such as sexual dominance, mating, feeding etc., however cetaceans often engage in a variety of behaviors 'just for fun'. Play also offers individuals, especially the young, time to practice behavior patterns that will prove useful later in life. Often though, aerial displays, erratic swimming and surfing are witnessed from individuals, irrespective of age.

Play underwater has also been documented in dolphin species playing with sponges and shells and making bubble rings. Dolphins have also been seen playing with other whales, creating social and playful interspecies interactions, where neither species is threatened. It is this element of play within social life which is a key component to understanding the intelligence of cetaceans.

Guidelines to Approaching Cetaceans

Different regions of the world have different guidelines and regulations with regards to approaching cetaceans, so please check first the regional rules with a local representative and report any violations. In general, there are certain rules that must be obeyed throughout the world to ensure safety and the minimum disruption to the cetaceans. Below Fig. 1 shows the regulated approach areas for cetaceans in the Azores.

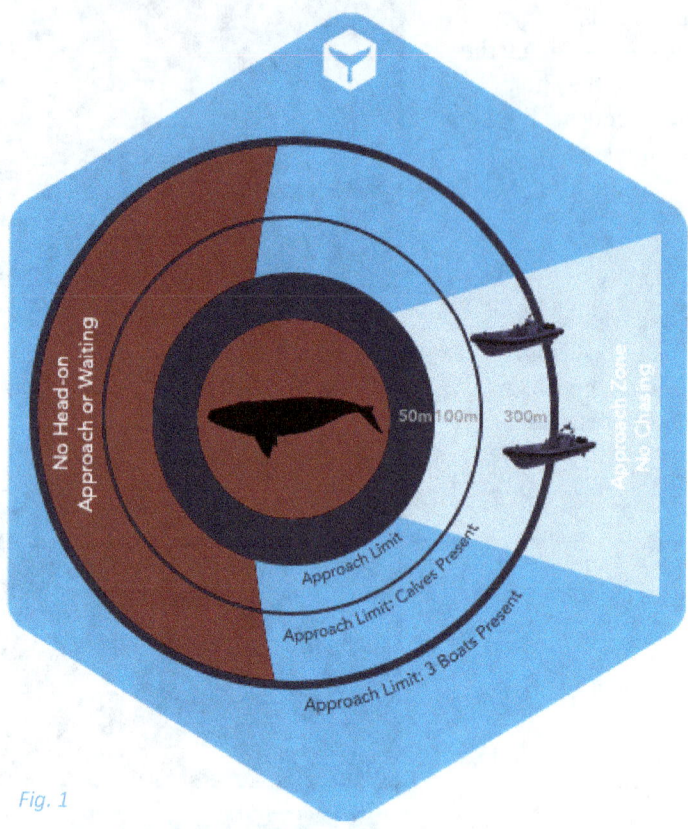

Fig. 1

In conjunction with the above the following rules are applied:

- Never purposely approach cetaceans closer than 50m
- Never purposely approach cetaceans head on or wait in their path
- Never purposely approach closer than 100m when adults are with calves
- Keep noise to a minimum that could disturb or attract cetaceans when in close proximity
- Never chase cetaceans
- Never separate or isolate cetaceans from the pod; especially calves from adults
- Never feed cetaceans
- Do not use sonar when you know cetaceans are in the vicinity
- Reduce speed and avoid any abrupt changes in course
- Stay on the offshore side of the cetaceans when they are travelling close to shore
- Do not scuba dive, swim and/or use underwater scooters inside the approaching area
- Do not pollute the ocean with any liquid or solid waste
- Night observations are forbidden, except for scientific purposes.

It is important that the captain of the vessel observing the whales is well versed in these rules so that they make the most sensitive approach possible. When entering the approach area, the vessel must be kept parallel and ever so slightly behind the animals, so that they have a clear line of sight 180 degrees ahead of them. Always maintain a steady slow speed, no greater than the speed of the cetaceans and keep an eye on their movements, avoiding any sudden changes in direction or speed. The engine should only be put into reverse in an emergency. Should the cetaceans approach the vessel closer than 50 m, the engine should be idled. All vessels in the approach area must have use of an engine, hence no vessel just under sail should enter.

At any point in time, should the cetaceans show signs of stress or appear aggravated, all vessels must increase their distance accordingly. It is strictly forbidden to intentionally approach sperm whale calves when they are alone at the surface. No vessels should be within a radius of 500m, should the cetaceans be immobilized or when females are in labor. The appropriate maritime authorities should be alerted when an injured or dead cetacean is spotted or with regards to any incident involving cetaceans.

Time of permanence inside the approaching area should be limited to around 30 minutes. Once observation has finished and sufficient scientific data recorded, the vessel must exit the approaching area from behind.

When there are more than one vessel in the approaching area, further regulations should apply. Firstly, at any one time, no more than three vessels should be in the approaching area, at a radius of 300 m (980 ft) around the cetaceans. Vessels inside the approaching area must be parallel to one another and positioned in a sector 60 degrees behind the pod. Any waiting vessels should be positioned behind, at a radius of 500 m (1650 ft) or more. All approach maneuvers must be coordinated by radio, led by the first vessel

to approach. The precedence of observation is in order of arrival to the approaching area or by proximity to cetaceans that surface at a distance within 500 m (1650 ft) from any given vessel.

Should any vessel host audio-visual crews, their objectives must be communicated to all other observation or research vessels operating in the same area. Any audio-visual operation must be carried out by professionals and supervised by local scientific personal with proven experience in cetology. At no point should the operation try to manipulate or influence the behavior of the cetaceans. Any violations should be reported.

Whale observation excursions in the Azores have benefited from the traditional knowledge passed down from whaling generations, who developed unique systems to spot whales and to approach them undetected. Although a part of history that was once reviled, it is now embraced as part of local culture and identity: one that has now manifested into the modern-day protection of the whale populations we have left.

The whale tourism on the Azores has grown at a controlled and sensitive pace, with seasonal restrictions and high levels of regulations and licensing. This being said, the industry is continually striving to do things better and ensure a harmonious relationship with these animals.

When partaking in any whale observation or research excursions the above protocols and regulations should be applied, in conjunction with any local laws. Please ensure that you choose and encourage others to only support excursions that commit to these best practices and work towards protecting the oceans. Remember that each disturbance builds upon the next. Whales need time and space to breathe between dives and for socializing and feeding. Make sure you are on a vessel that does not disrupt this.

The Role of the Lookout

In order to efficiently spot cetaceans, onshore lookouts are employed, a method that was first utilized during the whaling era. In the Azores, these are known as '*vigias*'. They are located on high points around the island, some using the same lookout stations used previously to hunt whales. The vigias utilize high powered binoculars to scan the oceans, looking for signs of whale blows, splashing and surface movements. They may also look for indirect signs, such as the congregations of large groups of birds- taking advantage of scraps in areas in which cetaceans may be feeding.

The vigias start early, often before the whaling research station is open. They radio in any information of sightings and provide directional information to the boat captains. There are 5 lookouts on Faial and Pico, who are able to cover the south coast of both islands, the north /north-east of Faial and the channel between Pico and São Jorge. On a clear day they are able to scan areas of the ocean stretching out 50 km (31 miles) from shore.

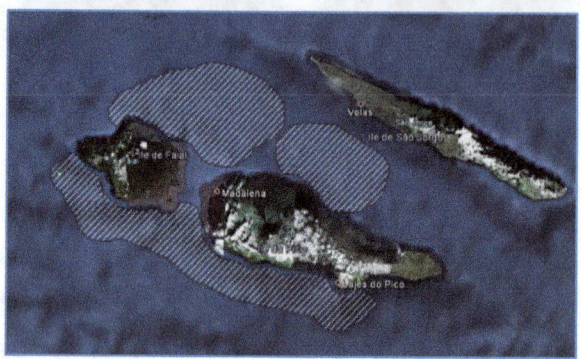

Observation Slates

When going out on cetacean observations, you will need to take a notepad or slate and record the following information:

Name:
Date:
Departure Time:
Arrival Time:

Species	Time of Sighting	GPS	Behavior	Markings	No. in Group	Observations

Identification Slates

Identification Slates

Chapter 2
Baleen Whales

"Baleen is a filter-feeding system which involves pushing vast amounts of water through keratinous plates."

Baleen whales are the largest animals to ever exist on the planet.

Chapter 2
Baleen Whales

Baleen Cetaceans of the Azores

Observations of odontocetes are best between the spring months of March and May, when these whales are migrating between the Arctic waters and the Azores. Recent research suggests that they use the mid-latitude areas such as the Azores as feeding grounds and navigational markers en route towards their summer feeding grounds. They time their arrival for the spring phytoplankton blooms in order to ensure a healthy supply of food intake on their journey.

There are 6 species of baleen whales that can be sighted in the Azores. In this section we will look further into each species and discover more about their biology and life history.

Blue Whale
Balaenoptera musculus

Blue whales are the largest animal ever to have lived on our planet, but little is known about their lives. Blue whales can reach a length of 24-28 m (78-92 ft), weighing around 100-150 tonnes (110 -165 tons), with females being larger than males (about 1.5 m/5 ft longer). The largest individual ever recorded was a female of 33 m (108 ft).

Currently, blue whales can be broken down into 5 recognized subspecies:

- *B. m. intermedia* – is the largest of the blue whales and is found in Antarctic waters. Commonly called the Antarctic/True blue whale.
- *B. m. musculus* – inhabits the north Atlantic and Pacific Oceans and is smaller than its Antarctic counterpart. Commonly called the Northern blue whale.
- *B. m. brevicauda* – colloquially known as the pygmy blue whale, it is the smallest species occurring in the sub-Antarctic zone of the southern Indian Ocean and the south-western Pacific Ocean. It is most abundant in waters off Australia, Madagascar and New Zealand.

- *B. m. indica* – assigned to the blue whales found in the northern Indian Ocean. Migratory movements of these whales are largely unknown but may be driven by oceanographic changes associated with monsoons.
- **Unnamed subspecies** – a small population of blue whales off Chile that are geographically, acoustically and genetically different from Antarctic blue whales. They migrate to lower latitude areas including the Galapagos Islands and the eastern tropical Pacific. They are intermediate in size, between the pygmy blue whales and the Antarctic blue whales, and are commonly referred to as the Chilean blue whale.

Distribution & Movements

Despite being reduced greatly due to whaling, the blue whale remains a cosmopolitan species separated into different populations from polar regions to the tropics. Migration patterns appear to be highly diverse, with food availability probably dictating its distribution for most of the year.

The majority of populations are migratory, feeding at higher latitudes during summer months and migrating towards the tropics or mid-latitudes in winter to breed. Acoustic and photo-identification data seems to indicate a wide-ranging use of the north Atlantic by blue whales, with migratory movements north in the spring and south in the autumn.

Social Organization & Behavior

We know relatively little about the social system of blue whales. They are usually seen alone or in small groups, and associations between individuals are short-lived. At good feeding grounds, concentrations of up to 50 have been observed.

However, considering that blue whales might hear each other and communicate over distances of tens of miles or more, one must be careful in presuming a lack of cohesion or group integrity.
Blue whales may raise their flukes high up in the air before diving; this is an individual characteristic, and it is mostly observed when an individual is relaxed. When foraging or feeding at depth, blue whales will normally dive for 8-15 min. and swim at 3-6 km/h (2-4 mph). When traveling, they can attain speeds of 5-30 km/h (3-19 mph) with bursts of up to 35 km/h (22 mph) when chased by boats, predators or interacting with other blue whales; but typical cruise speeds are around 2-4 km/h (1-2.5 mph).

Blue whales are the owners of the most powerful and deepest voice in the animal kingdom. The majority of vocalizations are low frequency or infrasonic sounds of 17-40 Hz (and up to 189 decibels), often described as pulses, grunts, groans and moans, undetectable by the human ear.

Blue whales vocalize regularly throughout the year, with peaks from mid-summer into winter months. The purpose of these powerful vocalizations is

unknown, but have been linked to communication, individual recognition, feeding, navigation, social organization and courtship. Distinct geographic variations in song types are helping scientists distinguish separate populations.

Diet & Feeding

Blue whales feed almost exclusively on euphausiids (krill) worldwide in areas of cold current upwellings. When they locate high concentrations of krill, they feed by lunging with their mouth wide open and gulping large mouthfuls of prey and water. A single blue whale may eat as much as 4 tonnes (4.4 tons) of krill and consume 3 million calories a day. During daylight hours blue whales commonly feed at depth, diving to at least 100 m (330 ft) into layers of krill abundance and staying submerged for up to 15 minutes. Surface feeding is most common during the evening, following the ascent

of their prey in the water column. When blue whales feed at the surface they often roll over on their sides and swim along with their mouths open.

Blue whales have the longest baleen of all the rorqual whales: 270-395 baleen plates on each side of the upper jaw, which can be 50-55 cm (1.6-1.8 ft) wide and up to 1 m (3.3 ft) long.

Life History

Little is known about mating behavior in blue whales and perhaps because of their powerful vocalizations, they do not seem to congregate on breeding grounds in the same manner as some other species, though female and calf encounters are commonly reported from different areas in late winter and spring (e.g. Gulf of California, Mexico, or the Azores).

Blue whales reach sexual maturity at around 8-10 years of age when females are 21-23m (69-75 ft) and males are 20-21 m (66-69 ft). Mating takes place starting in late autumn through winter. Depending on body condition and lactation period, female blue whales give birth every 2-3 years in winter, after a 10-12 month gestation period. The calves, which weigh 2-3 tonnes (2.2-3.3 tons) and measure 6-7 m (20-23 ft) at birth, are weaned when approximately 16 m (52 ft) long, at 6-8 months. Life span is thought to be at least 80-90 years for both sexes.

Conservation & Threats

Being the primary target species of modern whaling, the blue whale was hunted close to extinction throughout its range, and it is classified as 'Endangered' on the IUCN Red List of Threatened Species. Despite being globally protected since 1966, it is still an endangered species today, showing little sign of recovery in most of its range, with populations remaining at around 1-5% of their original population

The estimated global population is 5,000-15,000, with the north Atlantic population of blue whales estimated at 1000-2000.

Major threats today include overfishing and krill depletion, entanglement in fishing gear and ship strikes. Chemical pollution, long-term climate change and the increasing anthropogenic noise in areas of heavy ship traffic (e.g. Sri Lanka, California coast) may also have an impact on blue whale habitat and limit the recovery of their populations.

Identification

Body	Dorsal Fin	Blow
Enormous size	Small	Tall, vertical up to 9 m (30 ft) high
Blue-gray with mottled appearance	Set far back	
	Varies in shape – sickle-shaped, triangular or hooked	

Head	Underside	Fluke
Broad, flattened and U-shaped	Yellowish or mustard-colored, caused by diatoms which attach themselves to the whale's body (especially in polar waters)	Broad and triangular with a medium notch
Huge blowhole splashguard		Smooth trailing edges
Single longitudinal ridge on the rostrum		May show flukes when diving

Photo ID

Blue whales are approachable but sometimes difficult to keep track of, especially when feeding. They are inquisitive animals and so may approach the boat closely.

When diving they may or may not show their flukes and therefore this is not used as a reliable ID methodology.

Photograph the mottled pigmentation (ideally of both flanks) from the back of the head to the dorsal fin. Photos of scars and flukes may also be used for individual recognition.

ID Features

Photo of mottled region around the dorsal fin.

Fin Whale
Balaenoptera physalus

The fin whale is the second largest animal on Earth, but is also one of the fastest. Coined the 'greyhound of the ocean', it is able to reach speeds of 46 km/h (29 mph). Females are larger than males but on average fin whales can reach a length of 22.5-25 m (74-82 ft), weighing around 75-115 tonnes (82-126 tons). Northern hemisphere individuals are smaller than their southern hemisphere counterparts. Fin whales reach 95% of their maximum body size within 9-13 years of life.

Like most of the great whales, fin whale populations were heavily depleted in the 20th century, hunted almost to extinction.

Currently there are 3 recognized sub-species of fin whale:

- *B. p. physalus* - Northern fin whale
- *B. p. quoyi* - Southern fin whale
- *B. p. patachonica* - Pygmy fin whale

They are cosmopolitan animals, found worldwide, but with the highest densities in cool, temperate waters. They feed in higher latitudes in summer but aren't seen in waters close to ice packs near the north and south extremities. They then disperse to areas of the sub-tropics in winter for breeding.

Distribution & Movements

The distribution of fin whales tends to be determined by food. However, it is complex, with some populations showing migratory behaviors, moving to polar waters for feeding in summer and then again to lower latitudes to breed in the winter. These patterns however are not simple, and no wintering grounds are currently known in the North Atlantic. They are also semi-resident populations of the west coast of the US, the East China Sea and the west Mediterranean Sea.

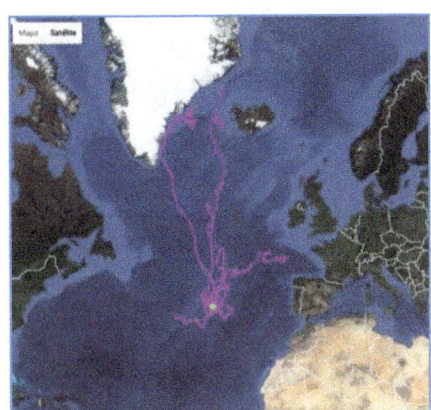

Fin whales avoid the extremities of the oceans at the poles and near the equator, and are found in waters deeper than 100-200 m (330-660 ft). They feed around the Azores and use the Mid-Atlantic Ridge as a navigation mark.

Social Organization & Behavior

Fin whales are generally observed singly or in small groups of 2-7. Long term bonds between individuals are rare and group members can often change.

They are relatively reserved, rarely demonstrating any surface behaviors such as breaches or fluking. This shouldn't be confused with shyness. They typically can be quite approachable and have been known to mix and socialize with many other cetaceans. There is some evidence of possible breeding between blue and fin whales, creating hybrids.

Male fin whales make a number of vocalizations including very loud low frequency grunts and moans as well as higher frequency songs made by infrasonic pulses that are downward sweeping from 16 to 40 hertz. Songs are composed of these single pulses for 1-2 seconds at fixed intervals and patterned sequences. These songs can be detected for hundreds of miles, reaching levels of up to 184-186 decibels. Song structure can vary between populations and variations are also found between seasons, suggesting a link between the song and breeding.

Diet & Feeding

When feeding, the fin whale blows around 5-7 times in quick succession, but when travelling, it will only blow once every minute or so.

Fin whales can spend several hours feeding per day, gulping up to 70 m3 (18,00 gallons) of water that sieves out through 350-390 overlapping, fringed baleen plates each side of the upper jaw. The baleen on the left-side has alternating bands of yellow-creamy and gray colors and on the right- side, the same except the front third which are a white/yellow color. The plates can measure up to 75 cm (2.46 ft) in length and 30 cm (1 ft) in width.

Fin whales are known to be opportunistic lunge feeders, and consume up to 1-2 tonnes (1.1-2.2 tons) of food per day of mostly krill and small schooling pelagic fish. One feeding technique they often use is to circle schools of fish, driving them into a ball, whilst rolling on to their sides and creating a 90 degree gape angle to engulf the massed prey. Lunges occur routinely on dives to 100-200 m (330-660 ft) for about 5-10 minutes or at the surface. Fin whales are also often seen in feeding aggregations with humpback and minke whales.

Life History

Adult males reach sexual maturity at 6-10 years old, depending on physical maturity. Females become sexually mature at around 7-8 years of age. Mating occurs in temperate waters during the winter and there is believed to be some competition between males, although this is less defined than many other mysticetes.

Calves are born after a gestation period of around 11 months, with peak birth rates in November/December in the Northern hemisphere and later in the springtime, in May-June, in the southern. Calving occurs every two or three years and calves are weaned after around about 6-7 months. Longevity is on average to 90 years old for both sexes.

Conservation & Threats

The travel speeds of fin whales, coupled with their preference for open ocean gave them some natural protection from early forms of whaling, however during the commercial whaling era after the decline of blue whales, the fin whale became a targeted species by the factory ships. Today fin whales are classified as 'Vulnerable' on the IUCN Red List. The current population of fin whales is believed to be less than 100,000 with the North Atlantic population estimated to be 47,300.

Identification

Body	Dorsal Fin	Blow
Large, sleek and narrow Dark-gray V-shaped chevrons on their back Prominent ridge on tailstock, giving it the name 'razorback'	Prominent, extending c.60cm (2ft) Sickle-shaped Rounded tip Leading edge rises at a shallow angle	Tall, vertical up to 9 m (30 ft) high

Head	Underside	Fluke
Asymmetrical head pigmentation – left-side gray and right-side white, known as a blaze Relatively flat Single prominent ridge on a sharply pointed rostrum Raised splashguard	Sides appear silvery, fading to a white/yellow color	Rarely raised Prominent median notch Smooth trailing edge

Photo ID

Upon surfacing, the fin whale will break the water with the upper-side of the rostrum. Key distinguishing features to look out for are the dorsal fin shape and position, as well as the dive sequence: the dorsal fin appears more rapidly after blowing than the blue whale, and they arch their tailstocks more than sei whales.

When diving the fluke is rarely shown, and so is not a reliable ID methodology.

ID Features

Photo of size and shape of dorsal fin.
Photo of any marks or scarring.
Photo of chevron pattern and the white pigmentation pattern on the right side of the head.

Sei Whale
Balaenoptera borealis

The sei whale gets its name from the Norwegian word for pollock, due to the simultaneous arrival of both predators looking to feed on abundant plankton in the summer months.

As the third largest whale species, sei whales measure around 15-20 m (49-66 ft) to a maximum of 19.5 m (64 ft), with females generally larger than males, weighing between 20-30 tonnes (22-33 tons). Also known as the 'lesser fin whale', sei whales are also fast swimmers, reaching speeds of up to 50 km/h (31 mph) in short bursts, which may make them the fastest marine mammals.

There are 2 recognized sub-species of sei whale, although there is currently insufficient evidence of genetic differences:

- *B. b. borealis* – northern sei whale
- *B. b. schlegelii* – southern sei whale (grows a little larger)

Distribution & Movements

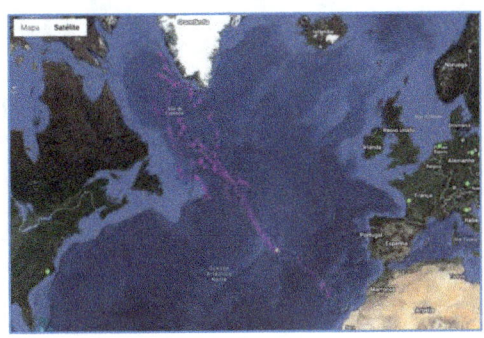

Sei whales live in all oceans, with distributions ranging from sub-tropical to sub-polar waters. They are cosmopolitan, with patchy oceanic distribution, but are most abundant in temperate, mid-latitude regions and are rarely sited in polar or tropical waters. Their migration patterns appear to be extensive and quite unpredictable, with limited distinction between feeding and breeding grounds. Whales observed in a region for a period of time may not return there for years, which is an unusual behavior for larger whales that usually show recurring patterns in their distribution.

Sei whales are generally found offshore, beyond the continental shelf, in predominantly deep waters, most likely following the bottom contours in search of food.

Social Organization & Behavior

Usually found travelling solo or in small pods of 2-6 individuals, sei whales have been known to congregate in large aggregations on productive feeding grounds. When in groups some social behaviors like chases and side swimming have been observed, but sei whales are rarely seen performing more acrobatic displays such as breaches.

Sei whales tend to sink below the surface with little arch in the back, which leaves a prominent fluke print on the surface. They will very rarely fluke up. They are known to occasionally follow a somewhat predictable line of trajectory, remaining visible, travelling below the surface for 20-30 seconds between breaths. However, surface behavior can also be quite erratic, with lots of directional changes. Sei whales maintain a normal swimming speed of 3-7 km/h (2-4 mph) but can swim at least 25 km/h (15.5 mph) in short, sharp bursts.

Diet & Feeding

Diet can vary and is adaptable, as sei whales feed opportunistically, depending on food abundance in the relative area. They primarily feed on copepods and euphausiids but will feed on small fish or squid, so long as these are shoaling in dense aggregations near the surface. Sei whales are predominantly skim feeders near the surface, but also gulp feed, consuming on average 900 kg (2000 lbs) per day. Having different feeding strategies allows sei whales to have a more generalist diet.

Life History

Sei whales reach sexual maturity at around 14 m (45ft) in length, between the ages of 5-15 years old. They usually give birth in mid-winter at lower latitudes after a gestation period of 10-12 months. Newborn calves measure about 4.5 m (15 ft) and

are weaned after 7 months on the feeding grounds, prior to the autumn migration. Longevity is thought to be at least 50-60 years or longer for both sexes.

Conservation & Threats

Listed as 'Endangered' on the IUCN Red List of Threatened Species. Although initially not a target of the whaling industry, the sei whale began to be exploited heavily only after the blue, fin and humpback whale stocks had been depleted. There has been very little research effort to assess the status of sei whales in all major oceans and there is evidence that some populations have not recovered.

Identification

Body	Dorsal Fin	Blow
Large, sleek and narrow Dark-gray; V-shaped chevrons on their back Prominent ridge on tailstock, giving it the name 'razorback'	Relatively tall, with leading edge rising at a steep angle Varies in shape – sickle-shaped, triangular or hooked, with tip which can be pointed or rounded	Tall, bushy blow of 3-4 m (10 –13 ft) Dorsal fin appears at the same time as the blowhole upon surfacing
Head	**Underside**	**Fluke**
Relatively flat with raised splashguard Single prominent ridge on a sharply pointed rostrum	Slightly lighter	Rarely raised Prominent median notch Smooth trailing edge

Photo ID

Notches and scarring on the dorsal fin and any other scarring or rake marks on the body.

When a sei whale begins a dive, it usually submerges by sinking quietly below the surface, often remaining only a few meters deep, leaving a series of swirls or tracks as it move its flukes, and rarely raising its tail flukes as it dives.

ID Features

Photo of the nicks and notches on the dorsal fin. Photo of any distinctive marks and scarring.

Humpback Whale
Megaptera novaeangliae

The humpback whale is the most famous of the great whales, known for their elaborate surface displays and courtship songs. They have a distinctive appearance with long narrow pectoral fins (*Megaptera* meaning 'big-winged') with scalloped forward edges and a number of knobs called tubercles on the head and lower jaw. The name 'humpback' comes from the curving motion of its rounded back while diving.

Adult humpback whales range between 14-17 m (46-56 ft), weighing between 20-30 tonnes (22-33 tons). Humpbacks travel immense distances during annual migrations. One individual migrated from feeding grounds in the Antarctic to American Samoa to breed- and back again- covering over 18,000 km (11,185 miles).

Distribution & Movements

Humpbacks reside throughout all the world's oceans. They travel great distances during migrations between summer feeding grounds in the higher latitudes to mating and calving areas in winter in tropical waters. However, there does exist one unique resident population found in the Arabian Sea, that remains in the tropical/subtropical waters all year due to monsoon driven food productivity in summer.

We can generally separate out the distribution of humpbacks into distinct global populations:

North Pacific

There are 4 known breeding populations that include:

- Mexican – breeding off the coast of Mexico and the Revillagigedo islands who feed mainly off California towards Alaska.
- Central American – breeding from southern Mexico down to Nicaragua that feed off the north-west of the US and British Columbia.
- Hawaiian – breeding around the islands and feeding either off British Columbia and southern Alaska or the Gulf of Alaska to the Bering Sea.
- Western North Pacific – breeding around Okinawa, Japan and the Philippines and feeding in the west Bering Sea and off the Aleutian Islands and Russian coast.

It is also believed that there is another breeding area somewhere else in the western North Pacific due to genetic anomalies of the population in this region.

North Atlantic

Feeding grounds range from the Gulf of Maine through the western coast of Greenland to Iceland and then on to the north of the Scandinavian nations. The majority of these individuals will then migrate to the West Indies to breed, however it is believed many of those feeding in more eastern regions breed in unidentified regions of the eastern tropical Atlantic, such as the Cape Verde islands, where several ID matches have been made.

The Azores archipelago and Bermuda are believed to be stopover points along migratory routes.

Southern Ocean

There are 7 known breeding populations that all migrate to areas around Antarctica to feed. These include:

- Atlantic coast of Brazil
- Pacific coast of Central/South America
- Oceania
- North-eastern Australia
- North-western Australia
- North-eastern Africa
- North-western Africa

As the global population of humpbacks has begun to recover, we have seen the return of former distributional ranges, including increased sightings in the Mediterranean and Baltic sea, as well the northernmost reaches of the Norwegian fjords.

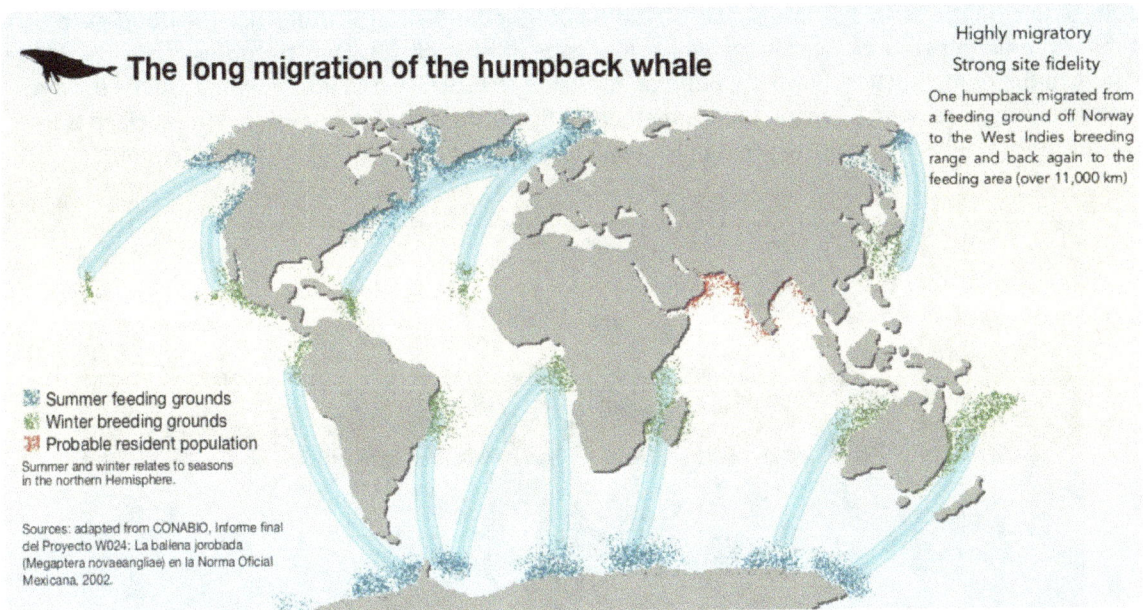

Social Organization & Behavior

The behaviors of the humpback whale have made it famous around the world. From their spectacular acrobatic surface displays to their melodic singing, more has been studied about the culture of these whales than any other. Throughout all seasons they can be seen breaching, lobtailing, pectoral slapping etc., for a variety of different reasons from parasite removal, communication, sexual prowess to sheer emotion and exuberance.

What has made humpbacks most famous however are their vocalizations. Both sexes make calls on

feeding grounds and on migration as part of social ordering and connection, however it is only the males that sing. Songs are most prevalent on breeding grounds and consist of a complex medley of around 36 different sound types – grunts, groans, moans, squeaks, etc. These can last anywhere between 5-30 minutes and are repeated over and over, sometimes for hours.

The songs of humpbacks from a particular area follow the same broad structure, however evidence has shown that they undergo improvisations and are influenced by others, hence males will often copy and start to sing new versions of a song. Songs have been transferred to different populations, however those in different parts of the ocean usually are singing very different 'tracks'. It is believed that singing is performed as part of sexual displays of dominance and courtship.

The humpback is also very social and friendly with different cetaceans and has been documented protecting other animals from killer whale attacks.

Diet & Feeding

Humpbacks feed on krill and a number of small schooling fish including herring, pollock, haddock and mackerel. They mainly gulp or lunge feed but also have a unique feeding strategy known as bubble-netting. Humpbacks will cooperate with one another to produce large circular bubble nets or clouds to encompass small fish into concentrated schools or bait balls. They then use their bodies to corral them into even tighter groups, flashing the white underside of their pectoral fins to confuse their prey and to herd them in the right direction. Once the bubbles appear at the surface, the group swims rapidly upwards with their mouths wide open to engulf the trapped prey. Sometimes the whale will slap the surface waters with its tail before bubble feeding, which is believed to create even more confusion to the school.

Some populations have also demonstrated other feeding techniques like pectoral herding to corral and disorientate prey; trap-feeding (when prey is more sparse) by opening their mouths at the surface and gently wafting their pectoral fins to scare fish into launching themselves inside their mouths or the 'trap'; and flick feeding by swimming around prey on the surface, slapping their tail to create a foam ring to encircle them, before emerging through the center to engulf the school.

Most foraging happens in the upper part of the water column, but humpbacks are capable of feeding at depths of at least 400m (1,312 ft). Little or no feeding occurs on winter breeding grounds, however, opportunistic feeding does occur during migrations.

Life History

Humpbacks reach sexual maturity between 5–10 years old. Females will produce a calf around every 2 years following a 12 month gestation period. Calves are weaned after about 1 year. Life span is thought to be over 50 years for both sexes.

Breeding is highly seasonal, with humpbacks typically preferring warm, shallow (<200 m/656 ft) waters that are surrounded by deep drop offs. Humpbacks have a high fidelity to return to their own birthplace to breed.

During breeding season, males exhibit competitive and aggressive behaviors around females, due to low sex ratio of up to 3:1 (due to the timing of migrations and the number of receptive females in a given year). Typically, males will escort or trail females, challenging each other through acrobatic displays including lunging, lobtailing, pectoral and tail slapping and breaching as well as high speed chases and ramming. It is also believed that song plays a key role in inducing sexual receptivity in females and as a factor in mate selection.

Conservation & Threats

The IUCN status of humpbacks is 'Least Concern' although some populations are still seriously depleted, especially around the USA. The species has demonstrated a great resurgence since the times of whaling, listed as 'Endangered' as recently only as 1988. Modern day threats of fishing gear entanglement, habitat destruction and overfishing of prey still persist to threaten the species.

Identification

HUMPBACK WHALE
Megaptera novaeangliae

Body	Dorsal Fin	Blow
Stocky	Low, stubby, hump shaped	Highly variable
Black or gray		
Long flippers		
Scarring from acorn barnacles		

Head	Underside	Fluke
Tubercles on the head and jaw	Variable amounts of white and mottling	Often flukes when diving
Acorn barnacle clusters		Variable pigmentation on underside

Photo ID

Humpbacks are inquisitive animals and so may approach the boat closely.

Often show their flukes. Flukes are white on the underside and have distinctive serrated trailing edges and pigmentations used for photo identification.

ID Features

Underside of tail fluke.

Minke Whale (Common)
Balaenoptera acutorostrata

In the Azores we observe the common minke whale, however there are two species:

- **B. acutorostrata** – Common (northern) minke whale
- **B. bonaerensis** – Antarctic (southern) minke whale

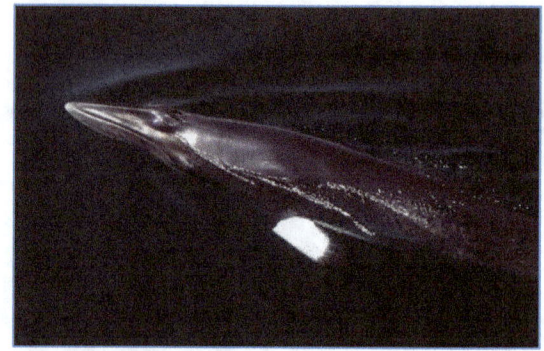

The taxonomy of the common minke can be further divided into 3 sub-species:

- **B. a. scammoni** - North Atlantic minke whale
- **B. a. acutorostrata** - North Pacific minke whale
- **Unnamed** - Dwarf minke whale found in the southern hemisphere, from the tropics to the polar waters of the southern oceans.

Minke whales are the smallest (second smallest of the baleen whales after the pygmy right whale), and most abundant of the rorquals. The name is believed to be derived from a Norwegian whaler 'Meincke' who mistook the smaller whale for a blue whale and hence the name stuck through the mockery of his mistake. The Latin name 'acutus' – sharp, 'rostrate' - beaked is in reference to their finely pointed rostrum which distinctively breaks the water first upon surfacing.

Minke whales can reach lengths of up to 10.7 m (35 ft) and typically weigh between 4-6 tonnes (4.4-6.6 tons).

Distribution & Movements

In the northern hemisphere the distribution of minke whales is considered cosmopolitan- with occurrences in polar to tropical waters, in both inshore and offshore areas. Seasonal migrations take place from the high latitudes in the summer to southern wintering grounds. In summer they are known to frequent more inshore waters in bays and fjords, but little is known about their breeding grounds in winter, suggesting these are mainly offshore areas. Some groups are not highly migratory and can be found year-round in areas of cooler waters, such as areas around the Canary Islands. They are rarely sighted in the Azores.

The dwarf minke is only found in the southern hemisphere, but very little is known about its distribution. They are more common in more temperate waters, frequently sighted around Australia, South Africa and the South American nations of Brazil, Argentina and Chile.

Social Organization & Behavior

Minke whales are usually solitary or found in small groups of 2 or 3. Observations of the distribution and segregation of minke whales in the northern hemisphere suggest a complex social structure.

They are commonly known to breach high out of the water and frequently spyhop, especially in icy areas. Their rapid surface swimming also creates a characteristic spray called a 'rooster-tail'. In some areas, such as north-eastern Australia, minke whales are very curious of boats.

Diet & Feeding

The prey choice of minke whales is seasonal and based on availability, feeding on a variety of schooling fish as well as krill. In order to corral fish into tight groupings, minke whales will swim in circles, ellipses and other patterns around schools of fish. The majority of feeding occurs at, or close to, the surface through lunging.

Life History

Minke whales reach sexual maturity between the ages of 5-8 years old. The gestation period is around 10-11 months. It is thought that calves are born every year and are weaned after 4-6 months. Longevity is thought to be between 30-50 years for both sexes.

Conservation & Threats

The IUCN status of minke whales is 'Least Concern'. Having been previously overlooked by whalers due to their small size, modern commercial whalers have since targeted the species for food- especially in Iceland, Norway and Japan.

Other threats include entanglement, noise pollution and vessel strikes.

Identification

Body	Dorsal Fin	Blow
Dark brown, gray, blackish Pale chevron on the back Prominent white bands on pectoral fins	Quite tall and falcate	Highly variable and indistinct

Head	Underside	Fluke
Longitudinal ridge on rostrum	White	Doesn't fluke when diving

Photo ID

Photo identification is not simple and can only be achieved by documenting distinct markings, scars or pigmentations.

Bryde's Whale
Balaenoptera edeni

Bryde's whales can reach around 15-16.5 m (49-54 ft) and range between 12-25 tonnes (13-28 tons). Unfortunately, they are one of the most undocumented and least known of the baleen whales. They are mainly found in warmer sub-tropical/tropical waters, greater than 16ºC. Their name is derived from a Norwegian named Johan Bryde (pronounced 'broodus') who was a pioneer of the whaling industry in South Africa.

In taxonomic terms, we often refer to the Bryde's whale complex, which lays reference to the on-going deliberation over sub-species divisions, since there is strong evidence of genetic and morphological differences warranting distinct species recognition. The current provisional classification of subspecies are as follows:

- *B. e. brydei* – Common Bryde's whale, found mainly offshore
- *B. e. edeni* – Eden's whale which is smaller and found in more coastal waters

The Omura's whale also used to be part of this 'complex', formally known as the 'pygmy Bryde's whale', as well as the Rice's whale, which was recently described as a distinct species.

Distribution & Movements

Bryde's whales are referred to as 'tropical whales' due to their preference for waters in lower latitudes between 40 ºN and 40 ºS, especially in areas of high productivity. There are no known extensive migration routes, however offshore individuals will make a more general movement between middle and lower latitudes between summer and winter respectively. In areas like the Gulf of Thailand and the Gulf of California, populations will remain year-round.

Social Organization & Behavior

Bryde's whales are mostly solitary or spend time in pairs, but can be found in aggregations of up to 20 individuals, in areas where food is concentrated. Research has suggested that Bryde's spend the majority of their time shallow, within the first 15 m (50 ft). They will sometimes breach, usually vertically, and occasionally many times in a row. Their behaviors can be quite erratic, with sudden changes of direction and irregular surfacing intervals.

Diet & Feeding

Diet consists of krill, red crabs, squid, other invertebrates and small schooling fish. Bryde's whales use a range of different feeding methods like skimming, lunging, creating bubble nets, or trap feeding.

Preference for prey has been demonstrated but they mostly feed based on prey availability.

Life History

Sexual maturity occurs between 6-11 years old. Coastal populations don't seem to show any seasonal restrictions in mating, but little is known beyond this. Females can give birth to a single calf every 2-3 years following a gestation period of 11-12 months. Calves are weaned after around 6 months. Bryde's whales of both sexes are thought to live between 40-50 years.

Conservation & Threats

The IUCN status of Bryde's whales is 'Least Concern'. They were not targeted by whalers due to their smaller sizes and absence from cooler waters, where hunts were concentrated. However, today they are hunted by Japanese vessels. Long term confusions between Bryde's whales and sei whales may have skewed historical data and this, combined with species complex, makes most analyses of population sizes data deficient.

Modern threats include pollution (specifically oil and plastics) and noise pollution, as well as vessel strikes and entanglement.

Identification

Body	Dorsal Fin	Blow
Dark, smoky gray body Slender pointed pectoral fins	Tall and falcate	Highly variable in height and can be dispersed or in a column Sometimes exhales under water
Head	**Underside**	**Fluke**
3 longitudinal ridges on rostrum	White or creamy	Doesn't fluke when diving No 'flukeprint' (unlike sei whales)

Photo ID

Photo identification of Bryde's whales may be possible from nicks and notches of the dorsal fin along with other marks and scarring.

END

Chapter 3
Sperm Whales

"Sperm whales have a spermaceti organ which is part of a system that produces the worlds most powerful echolocator."

Sperm whales are the largest of the Odontocetes and have the largest brain of any known living or extinct animal.

Chapter 3
Sperm Whales

Toothed Cetaceans of the Azores

The sperm whale is the largest of the odontocetes (toothed whales), equivalent in size to 4 elephants. Sperm whales frequent in deep waters (1000 m/3300 ft) throughout the year, but are best seen during spring and summer due to the better weather conditions at sea that enhance observations. Sperm whales are amongst the deepest diving cetaceans, diving to depths of up to 3000 m (9800 ft) while hunting prey such as the 13 m (43 ft) long giant squid.

The unique bottom topography within the Azores Archipelago, with deep water nearshore, is known to attract great numbers of sperm whales. Groups of females and immatures are most common, although 'bachelor' and adult males are also sighted.

Sperm whales are the key focus of our long-term studies in the Azores and so we will focus this chapter entirely on this species of toothed whale, discovering more about their biology and life history. Other toothed cetaceans shall be covered in Chapter 4.

Sperm Whale
Physeter macrocephalus

Sperm whales possess the largest brain on Earth. Their giant heads also contains the spermaceti organ (a waxy oil that early whalers thought was sperm) an extremely powerful, highly directional sonar system used for echolocation, hunting and communication. Sperm whales are the most sexually dimorphic cetaceans with adult females measuring about 10 m/33 ft, weighing 15-20 tonnes (16-22 tons). Males measure up to 18 m/59 ft and can weigh 35-50 tons (38-55 tons)

Sperm whales have a huge, squarish head (1/3 of the body length), often with scaring. The body is a dark brown to gray color with a lighter underside and wrinkled skin all over. The blow hole is orientated at 45 degrees, on the front left side of the head. In place of the dorsal fin, they have a thick, rounded and low hump.

10m female Pm skeleton in Horta's Whaling Museum

Sperm whales have a y-shaped lower jaw with 36-50 large conical teeth that are about 25 cm/10" long and weigh over 1 kg (2 lbs). Females have fewer, smaller teeth. Teeth in the upper jaw of both sexes rarely emerge. The presence of teeth in the lower jaw alone is typical of species that feed mainly or exclusively on cephalopods.

Distribution & Movements

Worldwide population estimation of sperm whales is between 300,000–400,000. They are found in all oceans of the world, but they are most abundant in offshore deep waters, although they may venture close to shore in the presence of deep-water geographical features such as canyons.

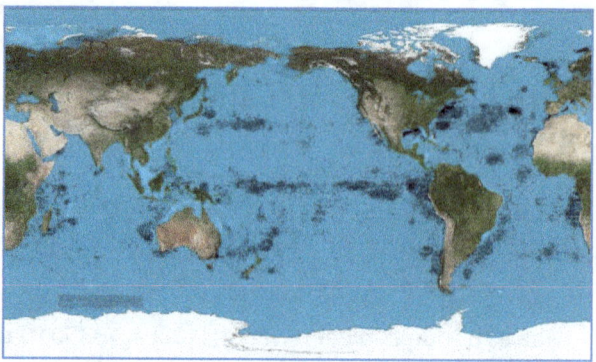

Distribution of sightings of sperm whales, based on the records of whalers and surveyors.

Distribution and movements vary considerably by sex and age, with little evidence of seasonal migrations. Where migrations are observed they are not very pronounced or consistent. Sperm whales typically share their ranges with, and may have relationships with, several thousand other individuals.

There is little known regarding the movements of mature males, as their habits are more complex than female sperm whales. Males show variable patterns of movement and larger home ranges than those of females. Some are more nomadic and roam more widely whilst some stay resident in small areas for long periods. What is known is that mature males live alone or in small groups, primarily in high latitudes, from about 40 °N/S to the ice edge, returning to tropical, warm-temperate waters to breed.

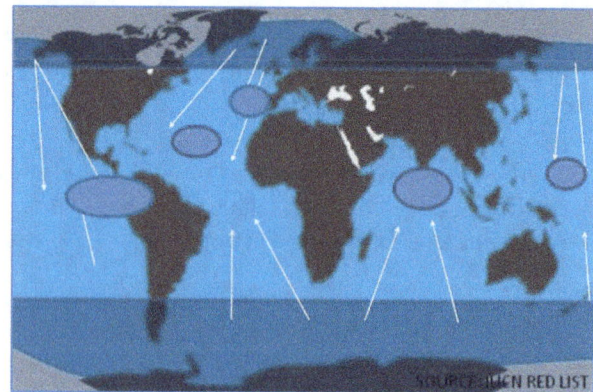

- **Females and males**
- **Only adult males**
- **Male migration towards lower latitude breeding grounds**
- **Females and immatures**

Females and their young (of both sexes) live in stable social units which generally remain in temperate or warmer waters all year round, and rarely venture beyond latitudes of 45 °N or 42 °S. Occasionally, females make long-distance movements of several thousand kilometers, but rarely move within ocean basins.

Social Organization & Behavior

Sperm whale social structure is based on matrilineal groups. Both male and female juveniles take an active role in the social lives of their units, but in their teens-twenties, with the onset of sexual maturity, the lives of the two sexes separate radically.

Females generally stay within their mother's 'social unit'. Unit size is variable, but averages about 10 individuals. Within social units there is a communal care for the young, with females babysitting and raising each other's calves. Members of social units are long-term companions. 'Groups' consist of two or more social units. These breeding groups of females with young of both sexes consist typically of 20 to 30 individuals, although there is much variation.

Sperm whales demonstrate cooperative vigilance and communal defense such as the marguerite formation, where individuals form a circle and all face inwards, creating a circle of large, powerful tails to protect the group or a weak/young individual.

Marguerite formation

The life of a male takes a very different turn when he leaves his mother's social unit. Although they are most often seen alone at the surface, nonbreeding males sometimes cluster with other males. Males in their teens and twenties can be found in loose 'bachelor groups', often consisting of individuals of about the same age. As the males age, their groups become smaller and gravitate towards higher latitudes. Males gradually become sexually mature in their teens, but do not seem to take much of an active role in breeding until their late twenties, when they make migrations from their cold-water feeding grounds to the warm-water habitat of the females. Older males tend to be solitary.

Sperm whales are gentle animals with very social behaviors. Female sperm whales live a highly social life, largely separate from the much more solitary males. Although there is much variation, groups of females and immatures spend approximately 75% of their time foraging and 25% socializing.

When foraging, sperm whales are seen at the surface in small clusters (1-3 individuals), moving relatively fast and consistently at an average speed of about 2-4km/h (1-2.5mph). They usually spread out to form a rank, which can be a kilometer or more long, and usually aligned perpendicular to the direction of travel.

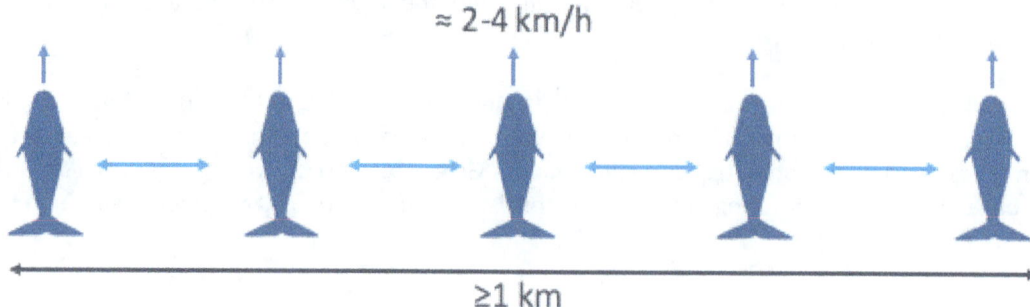

Sperm whales [...] :onds (females and immatures) to 17.5 seconds (large males). During their surface periods, sperm whales seem to actively cluster together, swimming along and frequently synchronizing their fluke-ups. If whales are disturbed during the surface period between dives, they may sidefluke, spyhop, or make shallow dives. The surface period ends with the raising of their flukes. Foraging dives last about 30-45 minutes. During most of their dives, the whales make trains of regularly spaced usual clicks, which are thought to be a form of searching echolocation.

Socializing often takes place in the afternoon. When females and immatures are socializing, clusters at the surface are larger and movements are slower and less directed. They can be extremely active (breaching, spyhopping, lobtailing, sidefluking) or they may lie still, apparently resting (logging).

only in the presence of a male, suggesting some kind of social interaction. Breaching may occur several times in a row. Aerial demonstrations like breaching and lobtailing may be a type of play, communication or used to remove parasites from the skin.

Large males socialize with females and immatures on the breeding grounds, but when they are at higher latitudes their social/resting periods seem reduced.

Sperm whales may be very vocal or remain silent. Vocalizations consist mainly of loud, broadband, directional clicks of between 5-25kH. Communication can occur many kilometers using a 'morse code-like' pattern of clicks called codas. These are the most distinctive communicative vocalizations of sperm whales made in social circumstances. Slow clicks (clangs), loud, ringing clicks repeated every 6-8 seconds are heard from large males, particularly on the breeding grounds.

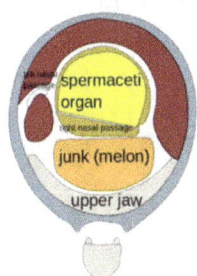
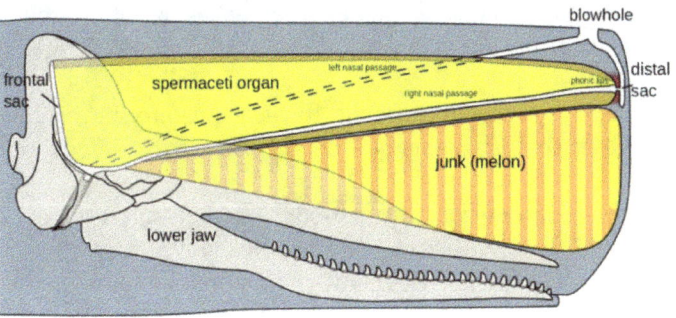

The spermaceti organ, which takes up 25-33% of the whale's body, is an extremely powerful, highly directional sonar system that is believed to produce echolocation clicks.

Diet & Feeding

Sperm whales mainly feed on squids, but also other deep-water species of invertebrates such as octopus, and fish such as sharks and rays. They can eat as much as a ton of food every day.

Feeding mainly takes place on or near the ocean bottom, foraging at extraordinary depths for their preferred prey. Sperm whales generally dive to about 300-800 m (980-2600 ft) for about 30 - 45 min, but occasionally descend to depths of 1000-2000 m (3300-6600 ft) for times in excess of 1 hour. Dives have even been recorded to 2800 m (9200 ft) for period of around two hours.

Calves and young sperm whales do not seem to make long foraging dives, and usually return quickly to the surface. Large males are known to make deeper and longer foraging dives than females.

Deep dives are usually preceded by a fluke-up and often have a U-shaped profile; dives are normally separated by periods of about 7-10 min spent breathing at the surface. When at depth, sperm whales seem to move horizontally at speeds similar to their speeds at the surface, and when ascending and descending – that is, about 4 km/h (65 m/min). When disturbed, sperm whales can make shallow dives, usually with maximum depths of 300 m (980 ft). These dives are not preceded by fluke-ups, and the descents do not seem to be vertical.

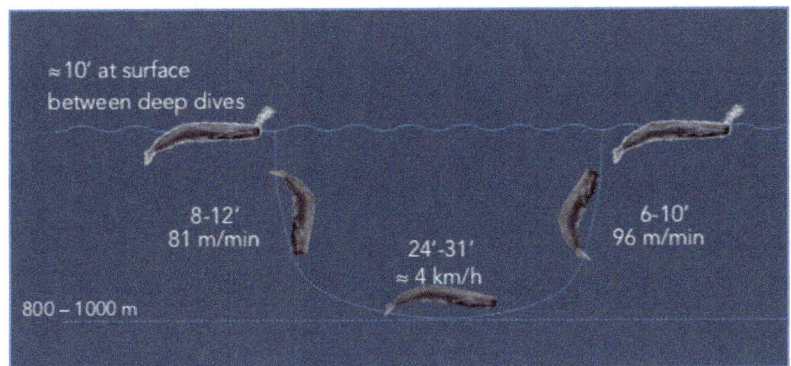

36-50 large conical teeth that may be 25 cm (10 inches) long and weigh over 1 kg (2.2 lbs)

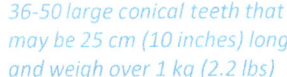

Life History

The breeding process of sperm whales is poorly understood. What is known to science is that large breeding males travel to warm waters to join female groups to breed and that they stay on low-latitude breeding grounds for at least a few months, roving between groups of females, searching for receptive mates, staying with each group for only a few hours at a time. When engaging with a group of females they produce extremely loud clicks, known as "clangs".

Female sperm whales first give birth at around 10 years old. Inter-birth intervals are generally about 4-6 years, but increase with age, so that pregnancy is very rare after age 40 (about one calf every 15 years).

Calves are born after a 15 month gestation period. A newborn sperm whale weighs about a 907 kg (1 ton) and is approximately 4 m (13 ft) long. They will suckle until about 2 years of age, although sporadic suckling may continue. There is also considerable evidence for communal suckling known as allosuckling. Weaning is gradual, and animals begin to feed on solid food at about 1-2 years of age. Individuals of both sexes can live over 90 years.

Female and immature sperm whales seem to actively change their dive schedules to improve the protection of calves, an apparent form of babysitting. Calves swim along at or near the surface above the foraging group, believed to be listening to the echolocation in order to keep up with their movements.

Conservation & Threats

Sperm whales are classified as Vulnerable on the IUCN Red List of Threatened Species. They have been heavily exploited by commercial whaling and although relatively abundant in most areas, some regional sub-populations have shown little signs of recovery (e.g. the Mediterranean). Sperm whales were targeted due to the high quality oils produced from the spermaceti organ- used for lighting (candles, lamp oils). Sperm whales also produce ambergris in their digestive system. As it ages this solid, waxy, substance acquires a sweet, earthy scent as it ages and has been highly valued by perfume makers as a fixative that allows the scent to endure much longer.

It is believed that the worldwide population of sperm whales was about 1,110,000 before whaling, which implies the industry caused about a 68% reduction in numbers. Sperm whales are still targeted in small numbers in Japan and Indonesia (by traditional hunting methods), however other evidence also suggests that sperm whales are being hunted illegally in some parts of the world.

Other threats facing sperm whales include entanglement in fishing gear, collision with large ships, noise pollution (acoustic trauma from military sonars, seismic exploration and other sources), chemical pollution and ingestion of marine debris.

From time to time, sperm whales, of all age and sex classes, mass strand on beaches. The stranded animals appear to be healthy, and most scientists believe these strandings may be caused by some combination of a failure of the sonar system to give clear information, external events such as poor weather, chemical pollution, noise pollution (acoustic trauma from military sonars, seismic exploration and other sources), and ingestion of marine debris.

Identification

The sperm whale can be identified by their broad flukes that are raised on diving. They dive for long periods and surface near the same place. They are often found motionless at the surface (logging).

Body	Dorsal Fin	Blow
Dark brown – gray body Wrinkled skin Knuckles from hump to flukes	Thick, rounded and low hump	Bushy blow oriented 45 degrees to the left and front Single S-shaped blowhole on left side near front of head
Head	**Underside**	**Fluke**
Squarish 1/3 body length	Lighter color	Triangular flukes with a prominent notch (up to 4m wide) Raised on deep dives

Photo ID

Show their flukes upon deep diving. Pigmentation and scarring on the underside and have distinctive serrated trailing edges used for photo identification.

ID Features

Underside of tail fluke.

END

Within these forms, orcas have vastly distinct ecological forms, which are referred to as 'ecotypes'. Ecotype is based on body size, coloration, habitat range, vocalization, prey preference, hunting techniques and social structure. Even though ranges often overlap, ecotypes do not associate with one another and therefore are genetically distinct. In total there are ten ecotypes:

Northern Hemisphere

- **Resident Killer Whale**
- **Bigg's Killer Whale**
- **Offshore Killer Whale**
- **Type 1 Eastern North Atlantic**
- **Type 2 Eastern North Atlantic**

Southern Hemisphere

- **Antarctic Type A Killer Whale**
- **Pack Ice Killer Whale (Large Type B)**
- **Gerlache Killer Whale (Small Type B)**
- **Ross Sea Killer Whale (Type C)**
- **Sub-Antarctic Killer Whale (Type D)**

As research continues it is likely that further ecotypes will emerge and some of the current ecotypes may gain separate sub-species or even species recognition.

Distribution & Movements

Orcas have worldwide distribution, found in open oceans, enclosed seas and coastal regions making them the most cosmopolitan of all the cetaceans. They are most abundant in higher latitudes around the Antarctic, Scandinavia and Alaska and have a preference for coastal over pelagic waters.

They are occasionally sighted in the Azores in the spring.

Social Organization & Behavior

Orcas tend to be very active at the surface, frequently lobtailing, breaching and spyhopping. When resting, a group will usually swim slowly and stay tightly packed together, synchronizing their movements but will become more spread out when feeding. Behavior, however, can differ greatly across the ecotypes.

Most of the known social organization of orcas is based on studies of residents in the North-East Pacific (NEP), from which we know the following: They form groups from just 3 to 20, depending on ecotype. They create strong family bonds with a matrilineal structure that consists of an older female, her offspring, her daughter's offspring and so on, up to around five generations. Members of these matrilines stay together for life. A pod consists of around 1-3 matrilines that remain with each other for the majority of the time. Pods can form communities with each other and those pods of resident ecotypes with similar vocalizations are said to be part of the same 'clan'.

It is unknown how much of the same social organization is found in other ecotypes, but it is likely that there is much more variety. For transient ecotypes the typical structure consists of a female and her offspring, but this social structure can be more fluid with offspring leaving the group for short periods of time, or even permanently.

Orcas are often highly vocal, communicating with each other during social interactions as well as during co-operative hunting in resident pods (transient ecotypes hunting marine mammals will remain silent whilst foraging in order to avoid detection). They produce three types of sound: clicks, whistles and pulses. Resident pods have their own distinct dialect that remains stable over time as they are culturally transmitted through matrilines to maintain group cohesion and family bonds.

Diet & Feeding

Diet is heavily dependent on ecotype and the cultural transmission of hunting mechanisms, based upon the available prey in the regions in which they frequent, that is passed on to offspring. Across all ecotypes, chosen prey is extremely diverse including other cetaceans, pinnipeds, sharks, seabirds, penguins and cephalopods as well as even terrestrial mammals like deer.

A range of specialized strategies have been developed within ecotypes to capture prey- including chasing, ramming and drowning of other whales, 'wave-hunting' to wash off prey resting on ice floes and even beaching themselves to catch seals close to the water's edge. Carousel feeding is another technique where they herd herring into tight balls and then stun the fish by slapping them with their tails. Sound is often used for cooperative attacks and to locate prey.

Life History

Orcas will reach sexual maturity at 12-15 years old. Mating is believed to take place when pods come together in order to diversify the genetic pool. Gestation is between 15-18 months with calving talking place every 3-8 years. Weaning occurs after 1-2 years. The lifespan is between 40 – 90 years, with females typically out-living males.

Conservation & Threats

The populations are mostly unknown with an IUCN status of Data Deficient, but some are endangered, specifically those in the Strait of Gibraltar and the southern resident population in the NW coast of the USA/British Columbia.

In the past orcas weren't targeted during commercial whaling, but persecution has taken place to kill orcas as a result of competition with fisheries. The most pressing issue for orcas in the modern era is captivity, where young individuals have been taken for the entertainment industry at places like SeaWorld. Other threats also include noise and chemical pollution, prey depletion and habitat degradation, especially for those that depend on ice to hunt.

Identification

Photo identification is possible with orca by capturing the size and shape of dorsal fin and saddle patch, along with any other scarring or notches.

False Orca
Pseudorca crassidens

The false orca is the second largest dolphin at 4.3-6 m (14-20 ft) and 1.2-2.2 tonnes (1.3-2.4 tons), with males growing larger than females. Its name refers to skull similarities observed in early carcass analysis rather than based on any behavioral or visual similarities to orcas.

False orcas have an entirely black or dark body, but lighter areas may occur ventrally. The pectoral fins are narrow, short and pointed with a distinctive bulge on the leading edge. The head is small and conical with a beak.

Distribution & Movements

Found worldwide, false orcas inhabit tropical to subtropical waters between 50 ºN to 50 ºS, usually found in deep, offshore areas and where deep oceans meet oceanic islands such as the Azores. They are best seen in the spring and summer months.

Little is known about the overall movements of the species beyond studies of specific populations, with some remaining in certain areas within a close proximity to the Hawaiian islands, and others seen travelling great distances along the Australian coast.

Social Organization & Behavior

False orcas are usually found in pods of around 20-100 individuals They are gregarious, with strong bonds and matrilineal family units. Smaller groups tend to join up on occasion to form larger aggregations of up to 400.

False orcas are highly social, fast swimmers that often breach and show lots of ebullience, and so very rarely are seen resting. They have been known to form groups with different species of dolphins, some forming long lasting associations. They are often inquisitive towards boats and may bow-ride or wake-ride.

The false orca is known to beach itself, sometimes forming mass strandings, the largest on record being 835 individuals in Argentina in 1946.

Diet & Feeding

False orcas primarily hunt fish and squid and have been often documented sharing prey (even with other species as well as documented attempts to share with humans when in the water with them). They mostly hunt near the surface, swimming at high speeds, often clearing the water. They also forage at depth, around 300-500 m (990 – 1650 ft) (but sometimes over 1000 m (3300 ft), diving for around 5 minutes.

Life History

Sexual maturity occurs in females between 8-11 years, with males maturing up to 8 years later. Gestation lasts for 12- 16 months. Calves are weaned at around 1.5-2 years, with calving occurring every 7 years. On average the lifespan across the sexes is thought to be around 60 years.

Conservation & Threats

The IUCN status of false orcas is Near Threatened. Modern day threats include intentional killing for meat or in retaliation for stealing fish from longlines, captivity, by-catch, prey depletion, human disturbance and habitat degradation.

Identification

Photo documentation of this species should include any distinctive marks and scarring on the dorsal fin.

Short-finned Pilot Whale
Globicephala macrorhynchus

Short-finned pilot whales are medium in size, reaching around 3.6-6.5 m (14-20 ft) and 1-4 tonnes (1.1-4.4 tons), with males becoming larger than females. They have a black to dark gray body (part of the blackfish grouping) with a bulbous head and a broad, falcate dorsal fin. Typically, they have light gray to white saddle patches, anchor shaped patches under the chin and blaze markings behind the eye. In the water it is extremely difficult to tell them apart from their close relatives, the long-finned pilot whale (*G. melas*), however they have shorter pectoral fins, less teeth and shorter beaks.

Distribution & Movements

This is a cosmopolitan species found in deeper tropical and warm/temperate waters. There are some known resident populations (e.g. Hawaii, Madeira) whilst others follow prey over long distances and travel based on seasonal temperature changes.

Social Organization & Behavior

Short-finned pilot whales are highly social with strong family bonds and a matrilineal family structure. Family groups can join up to form pods of around 20-90 individuals, where it is believed mating takes place. They can also often be found in aggregations with other species including larger whales such as humpbacks and sperm whales, other blackfish and dolphins.

Short-finned pilot whales are known to rest motionless at the surface but are slightly more active than long-finned pilot whales, performing surface behaviors such as breaching, lobtailing and spyhopping.

Diet & Feeding

Diet consists of mainly squid, but they may also prey upon octopus and mid to deep water fish on foraging dives that can exceed 1000 m (3300 ft) for up to around 15 minutes. Whilst foraging, pods of pilot whales can spread out over a large area of around 1 km (0.6 miles) wide.

Life History

Sexual maturity occurs first in females at about 9 years, with maturity in males occurring later, up to 17 years old. Females calve every 5-8 years. Gestation periods are 14-16 months and calves are weaned after at least 2 years. Longevity is at least 60 years in females and around 40 years for males.

Conservation & Threats

The IUCN status of short-finned pilot whales is Least Concern. They are susceptible to entanglement and are killed for interfering with fisheries. They are targets for meat in Japan, the Lesser Antilles in the Caribbean, and parts of southeast Asia. Vessel strikes, overfishing of prey species and other human disturbances also threaten the species.

Identification

Any photo documentation of this species is only achieved by recording distinctive marks and scarring, and the shape of the dorsal fin as well as the shape of any saddle patch that is present.

Bottlenose Dolphin
Tursiops truncatus

The bottlenose dolphin is probably the best known of all the cetaceans, popularized on television and in the entertainment and defense industries due to their high intelligence, trainability and high energy. They have small, robust bodies around 3-4 m (10-13 ft) and weigh 150-650 kg (330-1400 lbs), tending to be larger in the colder waters of their range.

There is a lot of taxonomic debate around the classification of the bottlenose dolphin. Two new species have been recently recognized; the Indo-Pacific bottlenose dolphin (*T. andicus*) and the Burrunean bottlenose dolphin (*T. australis*), although the latter is yet to be recognized by the Society of Marine Mammalogy's Committee on Taxonomy. Within *T. truncatus* there are currently 3 subspecies recognized:

- *T. t. truncatus* – the common bottlenose dolphin, found worldwide in tropical waters
- *T. t. gephyreus* – Lahille's bottlenose dolphin: a large form found along the coast of the western south Atlantic
- *T. t. ponticus* – Black Sea bottlenose dolphin

Much of the recent debate has been focused around the existence of two ecotypes in the North Atlantic. One a small, shallow water, coastal ecotype and the other a larger deep water, offshore ecotype. Genetic variation has been shown and so future species separation may be possible.

Distribution & Movements

Bottlenose dolphins are cosmopolitan, found in cold, temperate and tropical waters and present in the Azores all year round. Their range consists of a variety of different ocean habitats including open oceans, coastal waters, semi-enclosed seas, estuaries, bays and lagoons. Coastal populations tend to stay within a home range whereas offshore populations can travel long distances.

Social Organization & Behavior

Typically, offshore pods consist of 20-50 individuals but can sometimes be in their several hundreds. Coastal pods are smaller, at around 2-15 individuals. When traveling, pods can continually splinter into smaller units and reform. However, when resting, slow speeds are established and the pods gather tightly together. They are also often associated with other species of dolphin, false orca and humpback whales. Bottlenose dolphins show emotion towards each other through gentle body contact; they use play, aggression and sex as a way to socially interact.

Bottlenose dolphins show great exuberance in the water- surfing waves and lobtailing, alongside an array of other aerial acrobatics. When approaching a boat they will bow-ride and wake-ride at almost every opportunity.

Diet & Feeding

Bottlenose dolphins have a varied diet, and depth and duration of foraging dives varies greatly between populations, dependent on the location and type of prey. Echolocation and passive listening are used to search out both benthic and pelagic prey species.

They have been shown to be highly adaptable, demonstrating many specialized hunting techniques in different areas and for different prey. These include:

- Strand-feeding – driving prey onto muddy banks and temporarily breaching to retrieve their catch
- Hitting prey with their flukes (sometimes out of the water) to then consume whilst in a stunned state
- Assisted feeding with fisherman, driving fish into their nets so as to feed on the leftovers
- Herding fish into tight groups using high speed maneuvers and bubble blowing, then taking turns to charge through the school.
- Mud-ringing – beating the mud with their fluke, encircling prey who leap out of the circular plume into the mouths of other awaiting dolphins.

Life History

Sexual maturity amongst populations vary between 5-15 years old with females maturing first. Gestation lasts 1 year and weaning starts after about 1.5 years. Calving occurs every 3-6 years. The lifespan of bottlenose dolphins is believed to be between 40-50 years.

Conservation & Threats

Overall, the species are listed on the IUCN Red List as 'Least Concern', however there are some endangered populations as well as the endangered Black Sea subspecies. The species is hunted in these areas for consumption or as 'pest control'. Due to their high intelligence bottlenose dolphins are easily trained and, along with their ostensibly 'smiling' mouthline, have been favorited by the entertainment industry with over 1000 still held captive today.

Entanglement and pollution are other serious threats to bottlenose dolphins, as well as toxification from increased algal blooms.

Identification

Any photo documentation of this species is only achieved by recording distinctive marks and scarring on the dorsal fin and body.

Risso's Dolphin
Grampus griseus

Risso's have an extensively scarred body, large rounded head with no beak and a tall dorsal fin. When calves, they are dark gray to nearly black with whitening and scarring increasing with age due to intraspecific play or fighting. Because of this, individuals can look quite different with regards to coloring. They average 4 m (13 ft) in length and weigh around 300-500 kg (60-1100 lbs).

Distribution & Movements

Risso's are found mostly in warm, temperate and tropical waters between 15-20ºC, which results in some regional movements based on water temperature. They prefer deep waters but are also found in coastal regions and semi-enclosed seas, mainly where there are steep topographical features.
Risso's dolphins are present in the Azores all year round.

Social Organization & Behavior

Risso's are typically found in pods of 10-50 individuals but aggregations of thousands have been reported, as well as those travelling solo.

They can be quite active on the surface: flipper and tail slapping, spyhopping and breaching, as well as porpoising when on the move.

Diet & Feeding

Diet consists of mainly squid and octopus, with feeding taking place later in the day and at night. Due to lack of teeth on the upper jaw it is likely that Risso's suction feed.

Life History

Sexual maturity occurs first in females at 8-10 years and then in males from 10-12 years. The gestation period is 13-14 months. Males are weaned after 1 year. However, females can be weaned up to 2 years old. Longevity is believed to be 30-40 years for both sexes.

Conservation & Threats

The IUCN status of Risso's is 'Least Concern'. They are still hunted for food, bait or oil in some parts of the world as well as being vulnerable to entanglement, ocean noise and pollution.

Identification

Any photo documentation of this species is only achieved by recording distinctive marks and scarring on the dorsal fin as well as the unique patterns of body scarring.

Common Dolphin
Delphinus delphis

As its name suggests, the common dolphin is the most abundant cetacean in the world. They have a pale yellow and gray hourglass pattern on their flanks and have a prominent dorsal fin with a dark v-shaped cape underneath. Their average size is 1.8-2.4 m (6-7.9 ft) and they weigh 70-110 kg (155-240 lbs).

The current taxonomic consensus is that there is one common dolphin species, *D. delphis*, having once been split into two species, which included a long-beaked common dolphin, *D. capensis*. Evidence of common ancestry and lack of close relations between *D. capensis* populations caused the reclassification. There are currently 4 recognized subspecies:

- *D. d. delphis* – Common dolphin
- *D. d. bairdii* – Eastern North Pacific long-beaked common dolphin
- *D. d. tropicalis* - Indo-Pacific common dolphin
- *D. d. ponticus* – Black Sea common dolphin

Distribution & Movements

Common dolphins are found around the world in offshore warm, temperate to tropical waters of 18-22 ºC, most commonly found in areas of upwelling around sea mounts and ridges, as well as areas with steep sloping topography. In the North Atlantic, sightings are often associated with the Gulf Stream.

Social Organization & Behavior

Common dolphins are found in pods of 20-30 individuals but sometimes greater than 100. They are often associated with other cetaceans. They are highly sociable and aerially active, leaping and flipping high into the air. Upon re-entry to the water sometimes they make a large splash by slamming their bodies on the surface. When travelling they are usually doing so at high speed and are porpoising.

Common dolphins will often approach boats to bow and wake ride. They are also known to show altruistic behaviors to support injured members of a pod.

Diet & Feeding

Common dolphins feed on small fish and squid, mostly foraging in shallow waters, but are capable of diving between 200-300 m (650-1000 ft).

Life History

Females will become sexually mature first at around 8 years old whilst males, on average, mature at after 10 years. Gestation is around 11 months and calves are weaned after about 1-1.5 years. Longevity is believed to be around 30 years for both sexes.

Conservation & Threats

The IUCN list common dolphins as 'Least Concern' (note that they still recognize *D. capensis* as 'Data Deficient'). The greatest threat to this species is entanglement, often with one of the highest rates of bycatch. They are still taken for consumption and bait in some countries.

Identification

Any photo documentation of this species is only achieved by recording distinctive marks and scarring on the dorsal fin and body.

Striped Dolphin
Stenella coeruleoalba

Striped dolphins are named so due to the dark stripe along their flanks and their dark prominent dorsal fin. They also have a light gray spinal blaze sweeping back towards the dorsal fin. On average they measure 2.2-2.4 m (7-8f t) and weigh 90-150 kg (198-330 lbs).

Distribution & Movements

Striped dolphins are found around the world but have a preference for tropical to warm temperate waters (18-22 ºC) typically outside the continental shelf, in deep waters- in areas where ocean currents converge and in areas of upwelling. Found in the Azores year-round.

Social Organization & Behavior

Striped dolphins normally form large pods of between 10-500 individuals and often associate with the bottlenose dolphin.

They are very active, often performing chin slaps and a behavior known as roto-tailing, where they leap from the water and use their tails to create a circular motion. Striped dolphins often breach more than 6m (20ft) up in the air and swim very fast, porpoising, often upside down.

They avoid boats and will rarely approach.

Diet & Feeding

Striped dolphins feed on small fish and squid, on dives from 200-700 m (650-2300 ft), mostly at dusk. They have shown a preference for lanternfish in the oceanic North-East Atlantic.

Life History

Females become sexually mature first between 5-13 years, with males between 7-16 years, generally believed to be when they reach approximately 2 m (7 ft) in length. Gestation is believed to be about 1 year, with calves being born every 3-4 years. Weaning occurs around about 1-1.5 years. Their lifespan is around 60 years for both sexes.

Conservation & Threats

The IUCN status of striped dolphins is 'Least Concern' with the sub-population in the Mediterranean listed as 'Vulnerable' due to disease outbreaks from organochlorine pollutants. They are still hunted for food and bait in some parts of the world. Striped dolphins are also vulnerable to entanglement, especially in drift nets, and as by-catch in other untargeted fishing practices such as trawling.

Identification

Individual identification is not easy due to large pod sizes and natural shyness towards boats. The body coloration markings are not as clearly defined as other species. Any identification is achieved through distinctive marks and scarring, along with color variations.

Atlantic Spotted Dolphin
Stenella frontalis

This species is endemic to the tropical and temperate/warm waters of the Atlantic Ocean and are different to the pantropical spotted dolphin (*S. attenuate*) found in tropical waters worldwide. The appearance of individuals can differ quite drastically. However, most adults are heavily spotted with a lateral stripe, spinal blaze and falcate dorsal fin. Calves are born unspotted: spots beginning to appear between 2-6 years old, although some may remain without virtually any.

Atlantic spotted dolphins reach sizes between 1.7-2.3 m (5.6-7.5 ft) and weigh 90-150 kg (198-330 lbs).

Distribution & Movements

Found in the tropical to temperate waters of the Atlantic. The heavily spotted forms are usually found close to the continental shelf, with the more lightly spotted forms frequenting deeper oceanic waters in the high latitude areas of their range, such as areas like the Azores.

They are seasonally found in the Azores between June - October when sea surface temperatures are greater than 19 ºC.

Social Organization & Behavior

Atlantic spotted dolphins form pods of up to 100 individuals and can be found associating with other species, especially at feeding grounds.

They are very active and spend the majority of their time in surface waters less than 10 m (33 ft) deep.

They are known to love to bow ride, often seeking out boats from afar.

Diet & Feeding

Atlantic spotted dolphin prey upon fish and squid; sometimes coastal individuals will prey upon benthic invertebrates by digging their beaks into the sands. They usually dive around 10 m (30 ft) but can reach depths of c.60 m. (200 ft).

Life History

Sexual maturity occurs between the ages of 8-15 with some males reaching sexual maturity slightly later. The gestation period is about 1 year with calves born every 1-5 years. Weaning can be up to 5 years. Longevity of Atlantic spotted dolphins is thought to be greater than 30 years for both sexes.

Conservation & Threats

The IUCN status is 'Least Concern'. They are still hunted for food and bait in some parts of the world and face other threats such as entanglement in fishing gear such as gillnets and purse seines, as well as pollutants such as organochlorines.

Identification

Any photo documentation of this species is only achieved by recording distinctive spotted patterns, marks and scarring on the dorsal fin and body.

END

Chapter 5
Conservation & Research

"Research in the blue can be like research in the dark."

Through research we can find a proven solution to conservation needs.

Chapter 5
Research & Conservation

Cetacean Research

Most of the information we have on cetacean physiology, anatomy and reproduction has come from the examination of carcasses during the whaling times, and captive animals. However, since the 1980's, research on large whales has increased greatly, with scientists using less invasive and more efficient methods of data collection. In this chapter we are going to review some of the research techniques that are used at research stations.

Photo Identification

One of the main cetacean research techniques used across the world is long term photo identification and behavioral studies. Scientists and trained individuals alike can partake in the process by submitting photos to specialized and publicly accessible citizen science databases.

As described in the previous chapters, for different cetaceans there are species specific features including flukes, body pigmentations, natural markings, scarring and dorsal fin shapes that we use to document individuals.

Long-term photo identification studies give us detailed information on whales' movements, residency, migrations, population sizes, social structures, reproduction and behaviors. Computer programs allow us to more easily and quickly identify matches between different photographs from many different researchers and study areas. This helps scientists establish connections between feeding and breeding grounds of migratory species or determine whether there is any movement or exchange between neighboring populations.

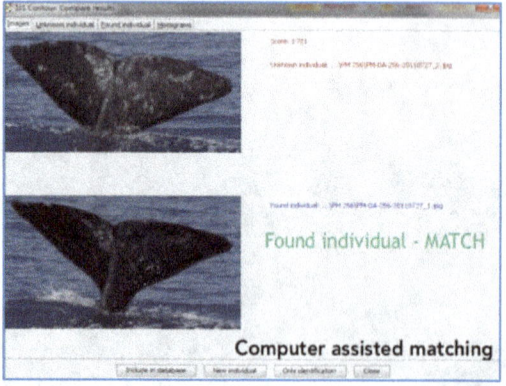
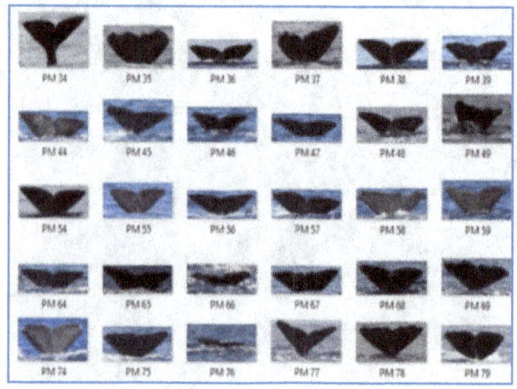

Photo identification is invaluable to whale research as it is inexpensive compared to other methods and, through courses like this one, can be made accessible to everyone- not just whale researchers.

How to Take ID Photos

When taking identification photos of cetaceans, the skills and setup required is more similar to sports photography rather than other types of wildlife photography.

It is recommended that you invest in an SLR or mirrorless camera base if you plan to actively partake in whale identification. However, point and shoot cameras will suffice in some circumstances. Your camera base will need to have a fast autofocus or servo-focus, which gives the ability to track and analyze movement and focus the image based on where it predicts the subject will be at a given point in time. The base will need a good drive speed (frames per

second rate) and burst or continuous shooting mode. If you would like a recommendation on the best and latest camera bases to fit your needs and budget, please ask your instructor.

Setting up your Camera

With an SLR or mirrorless camera you will be able to use detachable lenses, and we advise you to have a couple of them on your person when participating in whale observations. A wide-angle lens such as a 24-70 mm f/2.8 is great for closer shots. A long lens such as a telephoto 100-400 mm or 70-200 mm f/2.8 will allow you to take better shots at distance and gather enough detail to create quality photos to be used in photo identification. You also may be able to get away with using one lens with a varied focal length such as an 18-300 mm.

Always use a lens hood to stop glare and reduce salt spray. It is worth considering the weather and boat conditions before packing additional lenses, and if unsure which lens to use, please ask your instructor before going out on observation.

Shutter speed should be set fast: at least 1/400; but when set higher at 1/1500 or 1/2000 you will have a better chance of capturing unexpected surface behaviors. You can make this setting in shutter priority mode (S/Tv). If using aperture priority mode (A/Av) or manual mode (M) you will need to use ISO 400 or 800 to keep the shutter speed fast enough. Keep the aperture at around f/11 or f/16 when shooting from distance as you will need to react in a split second and so it is better to keep the depth of field large. For closer interactions you can get away with a shallow depth of field of around f/5.6.

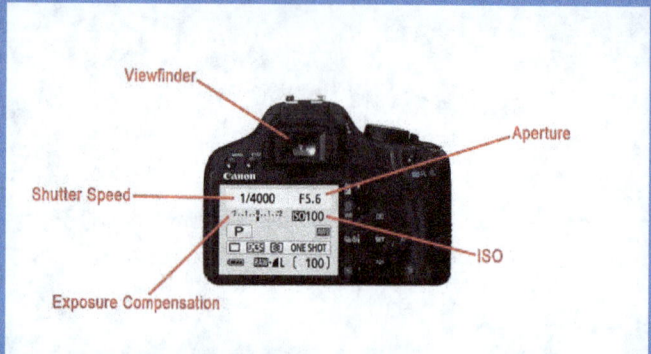

For white balance, set the camera to Auto; or if it is sunny, to Direct Sunlight. Similarly, if it is cloudy or overcast, set it to Cloudy.

Set your camera to autofocus and continuous shooting or burst mode. If the camera you are using lets you choose between Continuous High or Continuous Low framing rates and you are photographing a lot of action- or are afraid you will miss a great shot and want to capture as many frames as the camera will record- set it to Continuous High. Remember that when in continuous shooting mode you can keep your finger on the shutter button until it fills the buffer.

It is also advised to take spare batteries and memory cards since shooting in continuous mode will drain the battery quickly and fill cards quickly, especially if shooting in RAW.

One of the main benefits of capturing a photo as a RAW file is that the additional tonal and color data in the file offers more options, especially if exposure changes are needed. With a JPEG, white balance is applied by the camera, and there are fewer options to modify it in post-processing. If you are not comfortable with post-processing photos in RAW editors like Lightroom, use JPEGs.

Select memory cards with fast download speeds (90 MB per second or higher) to ensure that the camera's buffer will not suddenly stop mid-action.

Useful additional equipment may include a monopod for heavy lenses and a polarizing filter to reduce glare on sunny days.

In order to determine abundance of cetaceans within a given area, we use a method called 'mark and recapture'. Photo identification is the perfect method to use for 'marking' a whale since the method is completely benign. To 'recapture' a whale, it has to be subsequently re-sighted in the study area, photographed, and matched with the earlier photo.

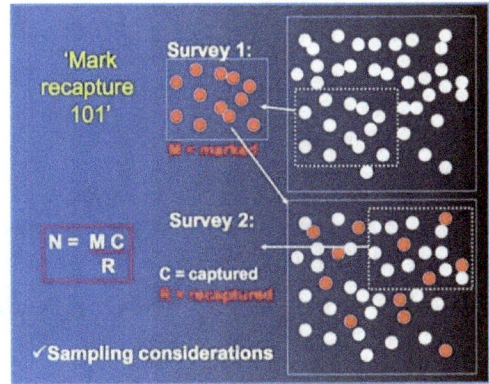

Examples of Photo Identification

Blue Whale Photo ID

- Can be individually identified by the mottled pattern on their flanks (unique to each whale).
- Perfect photo for ID, showing the dorsal fin and the mottled pattern on the side of the whale's body.
- Both flanks should be photographed; flukes may also be used for individual recognition.
- If possible, also photograph the tail stock (both sides).

Humpback Whale Photo ID

- Photographs of pigmentation patterns and scarring on the ventral surface of tail flukes, together with serration patterns along the trailing edge are used to individually identify humpback whales.
- Photographs should be captured from behind with both the trailing edges and underside visible, when the fluke is at its most vertical.

Sperm Whale Photo ID

- Sperm whales show their flukes before a deep dive.
- Each tail is unique. Distinctive scars, nicks and marks along the trailing edge of the flukes are used for individual identification.

Data Input

Most scientists, research stations and academic institutions will maintain their own catalogs of identification photos which they will exchange and collaborate with to find matches and produce scientific outputs.

They also apply computer vision algorithms such as Flukebook (www.flukebook.org) in order to identify and track individual whales and dolphins across hundreds of thousands of photos. Programs like this help researchers collaborate with each other across the globe and incorporate citizen scientists' contributions to the photo identification effort. Once digital "fingerprints" are made for individual cetaceans, the algorithms can rapidly scan photos against the image database. Outputs can help track populations, provide analysis of social structures and record migration patterns. All analysis of the algorithm is then reviewed by researchers.

Another algorithm that is focused more on supporting the citizen science efforts of non-scientists is Happy Whale (www.happywhale.com). They provide a user-friendly way for people to record encounters and provide valuable data to researchers.

Acoustics

Acoustic research can be used for individual identification, social population structure, migration and movements. During our observation trips we will use a hydrophone to better locate groups of cetaceans, especially during times of poor visibility and to determine the size of groups.

A hydrophone works as an underwater microphone, that when submerged in water converts sound pressure waves into electrical signals that can be heard as they are emitted and recorded.

You can apply a baffle on one side of the hydrophone that isolates the pressure waves, so that only the exposed side collects sound. This makes it a directional hydrophone, which can be used to determine which direction the sound originated from. Humans are unable to hear the infrasound and

ultrasound that some cetaceans produce. However, hydrophones can be set at different frequencies and record the sounds, which can then be later adjusted to a frequency humans can hear.

Tags and Time-Depth Recorders (TDRs)

During whale observations by boat, it is impossible to track movements of cetaceans for extended periods of time, especially when they are diving. By tagging whales, scientists are able to gather GPS positions of cetaceans, and track them to aid in the mapping of migrations and create computer simulations of behaviors. This data can be used to identify key breeding and feed sites, locate main migration routes, document responses to fluctuations in food supply, create diving profiles of foraging cetaceans as well as many other research possibilities.

Researchers apply tags or sensors that either fix to the whales' body with suction cups, or by temporarily anchoring them to the blubber under the skin.

TDRs and Digital Acoustic Recording tags (D-tags) are small, non-invasive data collection devices. VHR radio transmitters allow researchers to track cetaceans and recover the devices.

TDR
Time-Depth Recorders
Used to study diving behavior

11 TDRs in 2005, 2008 and 2009
(all performed foraging dives)

Dtags
Digital Acoustic Recording Tags
Record acoustics, 3D movement data and environmental parameters

11 Dtags in 2010
(7 performed foraging dives)

Satellite tags are used to study migratory patterns and daily behaviors by transmitting signals that produce telemetry data, revealing the cetacean position at regular intervals in order to provide insight into their long-range movements. Time and distance between data points also allows researchers to determine the different swimming speeds of migratory whales and infer possible feeding and behavioral patterns when they slow down. The downside of this research technique is that is highly specialized and expensive.

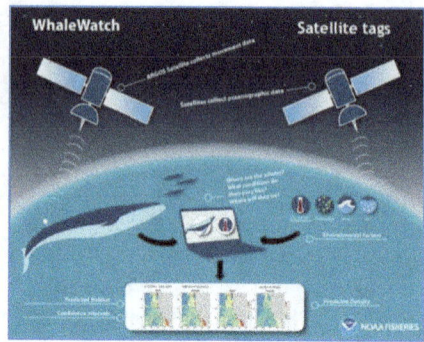

Genetic Sampling

Genetic analysis is becoming a more and more important tool to understand cetaceans and improve conservation efforts. Samples can be extracted easily from dead, stranded or entangled whales or from skin discarded from acrobatic surface behaviors. More technical sampling is taken using long poles tipped with scrubbers or darts to collect skin or excrement from swimming cetaceans.

Data taken can be used to determine sex as well as family lines between other samples of individuals within the same population and other neighboring groups. Analysis has helped to determine whether groups are discrete or isolated populations as well as helping to determine where new sub-species, species or eco-types exist.

Genetic sampling can be an invasive exercise and so is only performed under special licensing and by experienced researchers.

Line Transect

Researchers use transect lines to analyze encounter rates within a study area by navigating these lines over a defined time period and/or distance and recording cetacean observations. Lines are laid evenly spaced, parallel to one another or in zig-zags with the study area, so that vessels can systematically navigate and make recordings. By comparing samples over a set distance or time within the study areas, they can determine where the more populous areas are, as well as seasonal variation in encounter rates and habitat preferences- which can all help in the creation of marine protected areas and other conservation efforts. Transect lines can also be used to make abundance estimates.

Stranding Research

Up to 2000 cetaceans strand each year, usually involving odontocete species such as pilot whales, false killer whales and sperm whales. Research is being carried out to understand the causes of such events, which still remain mostly unknown. It is believed that some animals simply die at sea and are washed ashore, whilst other theories suggest solar storms affecting the earth's magnetism, and noise pollution (seismic surveys, oil prospection) affecting cetacean navigation.

Cetacean Conservation

Modern day cetacean conservation came to fruition in 1986 when, after tireless campaigning, the International Whaling Commission (IWC) imposed a moratorium on commercial whaling and set catch limits for all to zero. International trade of whale products is also prohibited under CITES (The Convention on the International Trade in Endangered Species).

By the time the moratorium was imposed, scientists estimated that worldwide whale populations had been reduced by 90% of their pre-whaling numbers, with some populations such as blue whales in the Southern Hemisphere declining by 99% and other populations like the North Atlantic gray whale already extinct.

Although the moratorium still stands today and has helped bring back some populations of cetaceans from the brink of extinction, many populations are still endangered and vulnerable. This is due to low reproduction rates as well as further and increasing threats from climate change, pollution and other anthropogenic sources. Blue whales of the Antarctic for example were reduced to less than 1% of their original numbers, and despite 40 years of complete protection, have yet to recover in any meaningful way. The Northeast Atlantic population of humpback whales is still as low as a few hundred and the Northern right whales are on the edge of extinction throughout their range. All small cetaceans were not protected in the moratorium, and the baiji river dolphin, which lived in the Yangzte River of China for about 20 million years, is now extinct as of 2007.

Modern day threats to whales and dolphins include:

- Commercial whaling
- By-catch
- Collision with ships
- Noise pollution
- Chemical pollution
- Marine debris
- Habitat degradation/disturbance
- Captivity
- Prey depletion
- Climate change

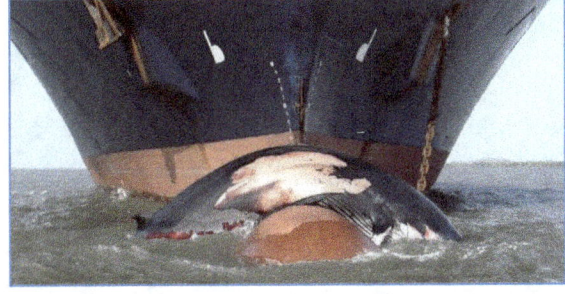
Ship collision killing a whale

As of the date of writing, 9 of the 16 great whale populations remain endangered or vulnerable:

North Atlantic Right Whale: **C. Endangered**
Western Pacific Gray Whale: **Endangered**
North Pacific Right Whale: **Endangered**
Blue Whale: **Endangered**
Fin Whale: **Endangered**
Sei Whale: **Endangered**
North Pacific Right Whale: **Endangered**
North Atlantic Right Whale: **Endangered**

Sperm Whale: Vulnerable
Humpback Whale: Least Concern
Common Minke Whale: Least Concern
Eastern Pacific Gray Whale: Least Concern
Southern Right Whale: Least Concern
Bowhead Whale: Least Concern
Bryde's Whale: **Data deficient**
Antarctic Minke Whale: **Data deficient**

The moratorium of 1986 unfortunately contained a loophole that allowed whaling nations to issue themselves permits to kill whales for research purposes under the guise that it is for scientific progress. Japan decided to exploit this and have continued commercial whaling. In 2019, Japan broke away from the IWC and resumed commercial whaling in its own waters for the first time in 30 years. Since it is no longer part of the IWC however, it is unable to legally hunt in other waters, thus ending its controversial expeditions to the Southern Ocean. Norway, Iceland and some other island nations have defied the moratorium from the offset, having never stopped commercial hunting of minke, sei and fin whales.

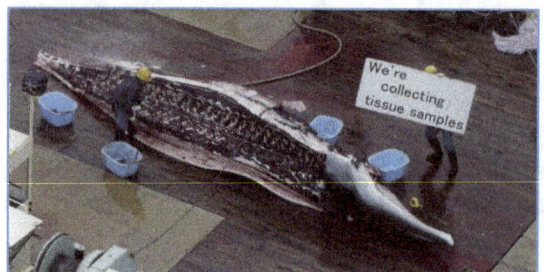
Greenpeace action against Japanese whaling

Minke whale killed by Japanese fishermen under 'by-catch' rule

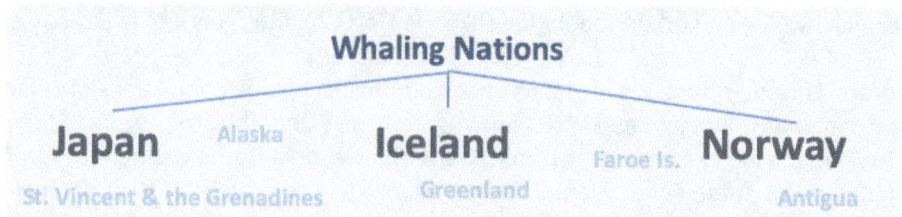

The Faroe Islands have made the news in particular: due to an annual hunt of pilot whales known as the 'grindadráp' that takes place between July and September, within 26 designated killing bays around the islands. The hunt kills up to around 1000 small cetaceans each year. Many conservation groups condemn the hunt, but it still remains legal under Danish rule as an important part of Faroese culture.

Increasing fishing of our oceans is threatening cetacean populations further due to entanglement, suffocation and underreporting of bycatch. Whales are also still being killed for meat and being served as tourist dishes in Scandinavian destinations. In 2009, 40% of the meat from whales slaughtered in Icelandic waters is eaten by tourists. Consumer habits need to be rectified in order to encourage better seafood choices from responsibly managed sources, caught using methods which minimize damage to the environment and other species. As of 2017, 11.4% of whale meat was eaten by tourists in Iceland due to successful campaigning.

The grindadráp on the Faroe Islands

The conservation effort must start with every one of us addressing our own consumer choices and educating ourselves on the issues at hand. Some ways to do this are to:

- Support organizations dedicated to the conservation and protection of cetaceans.
- Not support or visit any marine park or zoo that has captive dolphins and whales.
- Not participate in any captive swim-with dolphin programs.
- Voice your opinion about the killing of whales and ocean overexploitation: start a letter-writing campaign and/or sign existing petitions against whale hunting and wildlife destruction.
- Make good seafood choices – choose fish from responsibly managed sources, caught or formed using methods which minimize damage to the environment and other species (visit www.fishonline.org).
- Never dump anything in the sea or land. Plastics, fishing lines and hooks are serious threats to whales, dolphins, turtles, etc. Reduce, Reuse and Recycle.
- Support research stations and choose responsible whale watching organizations.

Citations

Carwardine, M., Camm, M., Robinson, R., & Llobet, T. (2020). Handbook of Whales, dolphins, and porpoises of the world. Princeton University Press.

Caitlin. (n.d.). Peering into the past: Cetacean evolution. WildWhales. Retrieved January 12, 2022, from https://wildwhales.org/2014/03/18/peering-into-the-past-the-evolution-of-cetaceans/

Magazine, S. (2010, December 1). How did whales evolve? Smithsonian.com. Retrieved January 15, 2022, from https://www.smithsonianmag.com/science-nature/how-did-whales-evolve-73276956/

Kooyman, Gerald & Castellini, Michael & Davis, R. (1981). Physiology of Diving in Marine Mammals. Annual review of physiology. 43. 343-56. 10.1146/annurev.ph.43.030181.002015.

A whale of an effect on ocean life: The Ecological and economic value of Cetaceans. Animal Welfare Institute. (n.d.). Retrieved February 14, 2022, from https://awionline.org/awi-quarterly/fall-2017/whale-effect-ocean-life-ecological-and-economic-value-cetaceans#:~:text=By%20the%20time%20a%20global,whales%20in%20the%20Southern%20Hemisphere%2C

Research methods used to study whales and Dolphins. Whale Watching Handbook. (2022, July 28). Retrieved February 12, 2022, from https://wwhandbook.iwc.int/en/industry-support/social-cost

Amorim, P., Perán, A. D., Pham, C. K., Juliano, M., Cardigos, F., Tempera, F., & Morato, T. (2017). Overview of the ocean climatology and its variability in the Azores region of the North Atlantic including environmental characteristics at the seabed. Frontiers in Marine Science, 4. https://doi.org/10.3389/fmars.2017.00056

Richardson, Anthony & Schoeman, David. (2004). Richardson AJ, Schoeman DS.. Climate impact on plankton ecosystems in the Northeast Atlantic. Science 305: 1609-1612. Science (New York, N.Y.). 305. 1609-12. 10.1126/science.1100958.

Roman, Joe & Mccarthy, James. (2010). The Whale Pump: Marine Mammals Enhance Primary Productivity in a Coastal Basin. PloS one. 5. e13255. 10.1371/journal.pone.0013255.

Li, Q., Liu, Y., Li, G., Wang, Z., Zheng, Z., Sun, Y., Lei, N., Li, Q., & Zhang, W. (2022). Review of the impact of whale fall on biodiversity in deep-sea ecosystems. Frontiers in Ecology and Evolution, 10. https://doi.org/10.3389/fevo.2022.885572

Welch, C. (2021, May 4). Groundbreaking effort launched to decode whale language. Animals. Retrieved April 20, 2022, from https://www.nationalgeographic.com/animals/article/scientists-plan-to-use-ai-to-try-to-decode-the-language-of-whales

Zandberg Lies, Lachlan Robert F., Lamoni Luca and Garland Ellen C. (2021) Global cultural evolutionary model of humpback whale song Phil. Trans. R. Soc. B3762020024220200242 http://doi.org/10.1098/rstb.2020.0242

Miller PJ, Johnson MP, Tyack PL. Sperm whale behaviour indicates the use of echolocation click buzzes "creaks" in prey capture. Proc Biol Sci. 2004 Nov 7;271(1554):2239-47. doi: 10.1098/rspb.2004.2863. PMID: 15539349; PMCID: PMC1691849.

Exploring our Fluid Earth. Activity: Whale Feeding Strategies | manoa.hawaii.edu/ExploringOurFluidEarth. (n.d.). Retrieved March 16, 2022, from https://manoa.hawaii.edu/exploringourfluidearth/biological/mammals/energy-acquisition/activity-whale-feeding-strategies#:~:text=Bowhead%20and%20right%20whales%20feed,as%20small%20crustaceans%20or%20fish.

Whale Hotspots of the World

Nuavut, Canada
- Narwal

In this massive, sparsely populated territory of northern Canada, the narwhal lives year-round
Best June-August

Iceland
- Sperm
- Blue
- Sei
- Humpback
- Minke

The North Atlantic is a good feeding ground; whale-watching trips run from the west or north of the country
Best May-September; June-July (blue)

Vancouver Island, Canada
- Orca
- Gray
- Humpback
- Minke

Some 20,000 gray whales pass the island's Pacific coast in spring, and it also has resident orca – the most researched pods in the world
March-April (gray); May-September (orca)

Baja California, Mexico
- Blue
- Gray
- Fin
- Humpback
- Bryde's
- Sperm
- Minke

Grays come to breed in San Ignacio Lagoon, on the Pacific coast; multiple species congregate in the food-rich Sea of Cortez
Best February-April

Dominica
- Sperm
- Humpback

Steep underwater drop-offs along the Nature Isle's west coast create sheltered bays – ideal places for sperm whales to breed and calve
Year-round but best November-March (sperm); January-April (humpback)

Churchill, Canada
- Beluga

More than 57,000 beluga congregate in the region in summer, in the warmer watch of the Churchill, after the ice breaks
Best July-August

Azores
- Sperm
- Blue
- Humpback
- Sei
- Fin

The remote archipelago sits in nutrient-rich waters; the seas sustain resident whale populations, while the islands are also visited by migrating species
Year-round (sperm); late March-early June (baleen whales)

Hervey Bay, Australia
- Humpback

The waters of the Great Sandy Marine Park, which is protected by Fraser Island, is a preferred place for humpbacks to rest during winter migration – especially with their young
Best mid July-October

Maui, Hawaii
- Humpback

Humpbacks make their annual winter migration through these islands
Best mid-December-April

Tonga
- Sperm
- Humpback

During the austral winter, humpbacks migrate from the South Pole to warmer Polynesian waters to mate and give birth
Best July-October

Sri Lanka
- Blue
- Sperm
- Bryde's
- Humpback

The southern tip nudges the depths of the continental shelf, favored by blues; nowhere else does the world's biggest creature swim so close to land
Best January-April

South Island, New Zealand
- Sperm
- Blue
- Southern Right
- Orca

Tectonic plates collide and ocean currents meet off Kaikoura, attracting an abundance of marine wildlife
Year-round (sperm); June-July (humpback); December- March (orca)

Western Cape, South Africa
- Southern Right
- Bryde's
- Humpback

Two oceans converge, resulting in a huge diversity of marine life; sheltered bays and warmer waters provide calving spots for migrating whales
Best July-Nov; May-Dec (humpback)